■ THE MOLECULES OF NATURE

THE ORGANIC CHEMISTRY
MONOGRAPH SERIES

Ronald Breslow, EDITOR
Columbia University

ORGANIC REACTION MECHANISMS, 1969
Ronald Breslow

THE MOLECULES OF NATURE, 1965
Third printing, with corrections, 1973
James B. Hendrickson

MODERN SYNTHETIC REACTIONS
Second edition, 1972
Herbert O. House

INTRODUCTION TO STEREOCHEMISTRY, 1965
Kurt Mislow

. THE MOLECULES
OF NATURE

A Survey of the Biosynthesis and Chemistry of Natural Products

JAMES B. HENDRICKSON BRANDEIS UNIVERSITY

1965
W. A. BENJAMIN, INC.
ADVANCED BOOK PROGRAM

Reading, Massachusetts

London • Amsterdam • Don Mills, Ontario • Sydney • Tokyo

THE MOLECULES OF NATURE: A SURVEY OF THE BIOSYNTHESIS
AND CHEMISTRY OF NATURAL PRODUCTS

First printing, 1965
Second printing, 1971
Third printing, with corrections, June 1973

International Standard Book Number: 33802-0
Library of Congress Catalog Card Number: 65-18900

Manufactured in the United States of America

ISBN 0-805-33800-4 (hardbound)
ISBN 0-805-33802-0 (paperback)
ABCDEFGHIJ-AL-7876543

■ EDITOR'S FOREWORD

UNDERGRADUATE EDUCATION in chemistry is in the midst of a major revolution. Sophisticated material, including extensive treatment of current research problems, is increasingly being introduced into college chemistry courses. In organic chemistry, this trend is apparent in the new "elementary" textbooks. However, it has become clear that a single text, no matter how sophisticated, is not the best medium for presenting glimpses of advanced material in addition to the necessary basic chemistry. A spirit of critical evaluation of the evidence is essential in an advanced presentation, while "basic" material must apparently be presented in a relatively dogmatic fashion.

Accordingly, we have instituted a series of short monographs intended as supplements to a first-year organic text; they may, of course, be used either concurrently or subsequently. It is our hope that teachers of beginning organic chemistry courses will supplement the usual text with one or more of these intermediate level monographs and that they find use in secondary courses as well. In general the books are designed to be read independently by the interested student and to lead him into the current research literature. It is hoped that they will serve their intended educational purpose and will help the student to recognize organic chemistry as the vital and exciting field it is.

We welcome any suggestions or comments about the series.

RONALD BRESLOW

New York, New York
December, 1964
v

■ INTRODUCTION

ALTHOUGH THE CHEMISTRY of natural products has always been a most important part of organic chemistry, it would be unrealistic not to admit that the beginning organic student today is introduced to very little of it: usually his lectures are consumed with the ordering of theory and concept, and his text offers only a cursory and fragmented picture in the last chapters. Defensible as this course may seem pedagogically, it is a great pity that the student misses so much of the excitement of organic chemistry, the intellectual stimulus of structural reasoning, and the esthetic elegance of synthesis because he does not discover the realm of natural products. Since part of the present difficulty arises from the lack of a book devoted to exploring the field, I have attempted in this volume to communicate some of my personal fascination with natural products to students of organic chemistry.

Only in the last few years, in fact, has it become possible to organize and unify the vast and diverse chemistry of natural products in a single-volume survey, since only recently has the essential simplicity of the pathways of natural synthesis of these compounds clearly emerged. These pathways of biogenesis are therefore taken here as a pattern for organizing a broad view of the heretofore dismayingly chaotic riches in this field.[1] Following the unwritten but widely accepted definition

---

[1] The present division of natural products into biogenetic groups is sometimes rough and arbitrary and often speculative, but it provides a convenient

of "natural products," the collection presented has been frankly limited to compounds primarily of chemical interest rather than biological significance and so excludes the amino acids, sugars, and pyrimidine bases and their polymers, the proteins, starches, and nucleic acids.[2]

Accordingly, the plan of the book is first to provide an understanding of biogenetic pathways adequate to group the natural molecules into families and encompass most of their structural variations; this provides an opportunity (Chapter 2) for a broad survey of structures. The major families then become the subjects of separate chapters in which the history of their study is encapsulated along with an outline of the major degradative tools which have been characteristic of that history. The bulk of these chapters (3–5) is devoted to a selection of particular examples chosen variously with regard to their utility as representative illustrations, their intellectual or esthetic appeal, or their pedagogical value for organic chemistry. The examples are also chosen to cover degradation and structure analysis, stereochemistry, synthesis, biosynthetic tracer work, and unusual reaction mechanisms.

Particular accounts are taken from the classical period as well as contemporary because both basically illustrate chemical reasoning and differ only in the tools applied. Finally, I hope to show in historical perspective something of the pattern of the symbiotic growth of natural products and organic chemistry with the aim of convincing the reader not only of the excitement but also of the past and present vigor and importance of the study of natural compounds.

This is an ambitious prospectus for a book of this size. In attempting to reveal the whole skeleton of natural products studies, we must cut close to the bone and dispense with much of the meat. This brevity has several consequences, however, for it allows the book to be read at several levels. Most simply, it can provide a brief scan of the size and shape of the field, a bit of its history and *modus operandi*, and a survey

organization for learning and orientation which is better than has been possible before. Our real biosynthetic evidence from tracer and enzyme studies is yet in its infancy and the reader is warned that some of the biosyntheses presented here could be discredited or seriously altered.

[2] The severe limitations on the length of this volume have also precluded the presentation of a number of interesting examples which do not fit smoothly into this brief biogenetic organization, viz., the natural porphyrins, pyrrol pigments, and chlorophylls, the penicillin family, xanthomattin, betanidin, gliotoxin, ergot and other modified peptides, vitamins, etc. Nevertheless, the great bulk of natural products are represented here.

of the molecular types involved. For this purpose its brevity should be welcomed and the requirements on the reader are easily met in a first-year course in organic chemistry. Digging deeper, however, the first-year student should find that the detailed examples are almost always within his grasp, for the chemistry has been restricted to well-known reactions, often briefly redescribed here as well, and the emphasis has been placed instead on the logic and reasoning involved.

Since the size of the book has required reducing each treatment to its essentials, the student should respond by giving the examples detailed consideration, often writing them out at greater length so as not to miss what is only implicit in the text—each brief account here represents only the outline of a major effort by a number of able chemists. Furthermore, brevity has often compelled simplifications in the original histories, the full flavor of which can best be appreciated only in the original or in extended reviews, to which the treatment here may then serve as a guide.

Finally, the really interested chemist is continuously invited throughout the book to solve problems of structural reasoning or reaction mechanism which are presented in the form only of data (adequate for solution) either with no solution included in the text or with the "answer" or outline of the reasoning deferred to a subsequent paragraph or page. Many of the famous problems from the natural product literature are presented in this way, some explicitly stated, others only implied.

Therefore, while I certainly cannot advocate this book as a text for the main course in organic chemistry, I do hope it may be served for dessert.

JAMES B. HENDRICKSON

Waltham, Massachusetts
June, 1965

■ CONTENTS

xi

■ THE MOLECULES OF NATURE

1

▪ PROLOGUE

THE WONDER AND CURIOSITY of man regarding his natural world are evident in his earliest records, and it is no surprise that his first explorations of the substances we now call natural products are lost in the mists of time. Crude aqueous extractions of flowers and certain plants and insects provided a number of pigments, such as indigo and alizarin, used in the ancient world for dyeing, and the recognition that mild heating of aromatic plants afforded perfumed distillates is at least as old. Primitive as well as civilized societies have always had an extensive tradition of the use of particular plants and plant preparations both for healing and killing. Thus the growth of chemical studies in the eighteenth and nineteenth centuries was inevitably deeply involved with increasingly more sophisticated probing into the nature of these traditional substances. Indeed it was the early recognition that these materials exhibited different properties and notably more complex and variable elemental compositions than did metal salts and rock materials that first led Berzelius in 1807 to define the separation into organic and inorganic chemistry from which we still suffer.

Organic chemistry was in the early nineteenth century exclusively the study of natural products, and was first intensively pursued in France by Dumas and his colleagues. They were unable to make much progress, however, despite many accurate analyses and pertinent chemical observations, since the essential distinction between organic and inorganic compounds, namely that between covalency and electrovalency, was slow to evolve. The dominant figure of the first quarter of the century was Berzelius, who wrote the first widely used text. While his careful analytical thinking went far toward clarifying electrovalent behavior, his insistence that all chemistry be unified under an electrochemical concept delayed the growth of organic theory, and it was not until 1859 that Couper and Kekulé independently proposed the constant tetravalency of carbon and wrote the first structural formulas. Butlerov first used the term "structure" in 1861, and in the following decade progress came with a rush.

The last half of the nineteenth century saw the rapid evolution of organic structural theory, one of the most astonishing and productive intellectual edifices man has ever produced. The development of the theory was closely tied to the formidable structural challenges presented by natural products. Van't Hoff's hypothesis of the spatial nature of molecules in 1876, for example, derived in large part from Pasteur's earlier observations on the stereoisomerism of the natural tartaric acids. In this period, while many of the main avenues of subsequent organic research sprang from natural products studies, at the same time organic chemistry was freeing itself from a reliance on natural compounds through the development of synthesis, the main impetus for which came from the challenges of natural compounds.

Most natural products have usually been obtained from plants or microorganisms since the practical difficulties in extracting them from animals are much greater in most cases. The folk medicines, the perfumes, and the coloring matters of tradition have generally stimulated the interest guiding the selection of materials to be examined, and so a general survey of botanical families in terms of their chemical constituents has not been made. It is probably fair to say that less than 5 per cent of all plant species have been explored.

The common procedure for obtaining natural products involves extracting the dried and ground plant material with a suitable solvent ($CHCl_3$ or CH_3OH, often) and evaporating this to a gum which must be separated into its components. Since the gum invariably contains an enormous variety of compounds, the natural products actually obtained pure for study have in the past usually been simply those which, by virtue of sheer predominance or differential solubility, were crystallizable from the mixture. Distillation sufficed in the case of liquids—and consequently they suffered from a lack of reasonable criteria of homogeneity—but most natural products have been crystalline solids. The natural organic acids and, especially, bases (alkaloids) predominated in classical chemistry as they were the more easily separated from the crude plant extracts by basic or acidic aqueous extraction.

Most commonly, the natural products so isolated are given names derived from the species name of the parent plant (cf. narcissidine from *Narcissus poeticus*). On some occasions other sources, such as the physiological action of the compound, have served, as in morphine, emetine, vomicine, pukateine, and putrescine.

In modern times the use of systematic column and gas chromatography has allowed separation from crude extracts of many compounds hitherto unavailable in pure form so that it is now not uncommon to characterize 30–50 separate pure natural products from a single plant species that might have yielded only two or three in 1900. This, of course, implies that many species would benefit from reexamination by modern techniques and in some cases this has already been very revealing.

The first problem posed by a natural compound is the elucidation of its structure. In a rough way the classical approach may be divided into four phases: determination of the functional groups; delineation of the carbon skeleton and location of the groups on it; clarification of the stereochemistry; and synthesis of the molecule in confirmation. Historically, the oldest procedure involved attack on the compound by chemical degradation, synthesis of the simpler compounds so derived, and reconstruction of the original molecule by inference. The next major step was the recognition of families of like structures, now understood to be the result of common biogenetic pathways. This step could be taken with confidence only after a fairly considerable backlog of known structures had already been amassed. Once taken, it permitted important analytical shortcuts, for a judicious assumption of structure often allowed chemical interrelation of an unknown with a previously elucidated compound of the same family.

The most important modern development in structural analysis has been the extensive use of physical measurements on the resting molecule. Such observations have the dual advantage of obviating unexpected changes in the molecule on chemical activation and of requiring only minute quantities of material (a milligram or less) and returning this unchanged. The earliest physical measurements were melting and boiling points and optical rotation, which were generally used for identification. A system for correlating these physical data with structural entities was eagerly sought from the start, but met with little success. The nineteenth-century literature also abounds with careful goniometric crystal measurements in the apparent though abortive hope that they would provide structural clues if only enough data could be amassed to allow correlations to be revealed.

In the 1930's ultraviolet measurements (UV) were beginning to be utilized and their correlation with the presence of conjugated unsaturation recognized, as in the succinct and successful Woodward's Rules (1941). The development of infrared (IR) spectroscopy followed in the 1940's—spurred by the wartime penicillin structure problem— and soon afforded clear indications of the presence or absence of a wide variety of functional groups, and of the size of carbonyl-containing rings. In the 1950's came the growth of nuclear magnetic resonance (NMR) spectroscopy with its unique and valuable property of discerning the environments of separate hydrogens in the molecule. This has been especially important since hydrogen is the most common element in organic chemistry. The 1960's are witness to the rise of the potentially very illuminating mass spectrum, in which molecules are fragmented by an electron beam, and the fragment masses ascertained.

The amplification of molecular rotation into the much more revealing optical rotatory dispersion (ORD) measurements, and the developments of the theoretical understanding and structural consequences of acidity and basicity measurements (pK) are the other major modern instrumental techniques. These techniques have largely usurped the primary place once held by chemical degradation in elucidating unknown structures, and have made the job far easier, quicker, and amenable to success with far smaller amounts of substance.

The logic of structure elucidation is not easily codified, but the common chemical tools can be briefly summarized. The fundamental fact of major importance is the empirical formula of the compound. Coupled with the particular expectation of empirical change for a given reaction, and the empirical composition of the product, this can afford considerable information, and any reading of the examples in Chapters 3–5 is incomplete without a careful consideration of the empirical compositions and their changes at each stage.

The first information afforded by the elemental analysis is the extent of functionality, from the number and kind of heteroatoms. When these have all been accounted for by functional group analysis, the formula then allows a determination of the number of rings from the sites of unsaturation implicit in it. In order to do this, the number of double and triple bonds must first be determined, usually by catalytic hydrogenation, while aromatic rings reveal themselves in a number of ways, such as the $FeCl_3$ color of phenols, the nitrosation of amines, or the UV spectrum. Quantitative determinations have also been widely practised to' ascertain the number of methyls attached to oxygen, nitrogen, or carbon. In the first two (Zeisel determinations) heating the compound with HI liberates methyl iodide, which is quantitatively estimated. In the C-methyl, or Kuhn-Roth, determination hot chromate oxidation is used to destroy all the structure except the CH_3—C– moieties, which are burned down to the relatively impervious acetic acid and this distilled and titrated (C—C_2H_5 can be determined by careful isolation of propionic acid also). In current practice methyl attachments are commonly recognized by their distinctive three-proton NMR peaks. Active hydrogen determinations (–OH, NH–) are traditionally made by reaction with methyl Grignard (CH_3MgI) and manometric measurement of the methane gas generated (Zere-witinoff determination), but deuterium exchange (with D_2O) is now widely favored as easier and more reliable.

Other functionalities are recognizable through a variety of chemical as well as physical means (cf. IR and UV spectra, etc.). A common test for primary and secondary alcohols consists in their conversion to acetates with acetic anhydride—tertiary alcohols being distinguished by their unreactivity—and the number of acetoxyl groups in the product counted by empirical composition or quantitative hydrolysis and titration of acetic acid. These alcoholic groups are also often deduced from their oxidation (CrO_3) to aldehydes and ketones. Ketones themselves are commonly identified as oximes, phenylhydrazones, and similar derivatives, and the presence of methylene adjacent to ketone by formation of a benzylidene derivative through an aldol reaction with benzaldehyde in the presence of base. Carboxylic acid derivatives are recognized by hydrolysis to the free acid, characterized by its acidity and titration for its equivalent weight.

The development of a number of reagents both mild and specific for certain functionalities has made functional group identification much more versatile. Typical of these are reductions with $LiAlH_4$ and

$NaBH_4$, MnO_2 oxidation of allylic alcohols, and HIO_4 cleavage of glycols. In all these reactions one depends on the reliability of expectation. This is certainly not always dependable and there are two simple checks available. First, the change in empirical composition must correspond to expectation for the reaction used, and second, wherever possible, the reaction product should be reconverted to the starting material to ensure that no unexpected (cf. skeletal) molecular change was involved.

More difficult than functional group analysis is the discernment of the molecular skeleton. Here two kinds of degradative tools have been used. In the first type the molecule is simplified by chemical reaction to a closely related stable aromatic skeleton, in many cases containing most or all of the original skeletal atoms. This simpler aromatic compound, often recognizable by a distinctive UV spectrum, may then be synthesized in confirmation. In this class of procedures lie zinc dust distillation; KOH fusion; and dehydrogenations by pyrolysis with sulfur, selenium, or palladium (see p. 106).

In the second type the molecule is cleaved into smaller fragments—each one much simpler and separately identifiable—by chemical attack on a vulnerable functional group or groups. In this category lie double-bond cleavage by O_3 or cold $KMnO_4$ oxidations, as well as the useful hot $KMnO_4$ which oxidizes all the side chains of aromatic rings (not bearing OH, SH, or NH groups) to carboxyl groups, so that identification of the product benzoic acid indicates substituent orientation on the original aromatic ring. The Hofmann degradation of amines, widely used in classical alkaloid studies, involves exhaustive methylation of the amino group to a quaternary salt and pyrolysis of its hydroxide to generate an olefin (the "methine"). Successive Hofmann degradations are carried out until the amino nitrogen is lost as trimethylamine, and the olefinic groups of the methines are available for ozonolytic (or $KMnO_4$) cleavages. Apart from occasional use of oxidants other than O_3 and $KMnO_4$, the other general fragmenting procedure simply involves hydrolysis with acid or base, wherein esters and amides, β-dicarbonyl compounds, and other labile functionalities are cleaved.

Once the skeleton and its functional groups are assigned there remains the problem of stereochemical delineation, since many natural products have a number of asymmetric centers and are optically active. Absolute configurations are demonstrated either by isolation of an asymmetric center intact in a degradative fragment and its interrelation with a known asymmetric compound, via a sequence of chemical interconversions; or by the interpretation of the ORD of a ketonic derivative, since these are now well correlated with absolute configuration, largely through the work of Djerassi.

Relative configurations at different centers are commonly assigned in one of two ways. An asymmetric center may be equilibrated and the nature of the thermodynamically more stable epimer deduced by conformational analysis. This is usually done with asymmetric hydrogen α- to a ketone or via aluminum alkoxide oxidation-reduction equilibration of secondary alcohols. Secondly, two asymmetric centers may often be assigned *cis* or *trans* relative orientations by observing a cyclization reaction between them, as in the lactonization of *cis*-related carboxyl and hydroxyl groups on a ring system. Since natural products run to structural families, however, the fastest way to determine both skeleton and stereochemistry lies in the conversion of an unknown to a previously elucidated compound.

Finally, quite apart from the rapidly expanding sophistication of chemical and physical methods in structure determination, the historically and operationally independent approach of X-ray crystallography has in recent years been so revolutionized through automatic data-gathering and computer treatment that it is frequently possible to provide a complete structure for a complex molecule in only a few months time. The future will undoubtedly see competition between the X-ray approach and the traditional physico-chemical operation of the chemist; and although it is by no means clear that X-rays will take the upper hand for speed, it is certain that structure elucidation in the future will be a relatively simple process either way, opening up the prospect of a great expansion in our knowledge of the structures of natural compounds. A corollary effect may well be the channeling of relatively more effort into synthesis, the tracing and simulation of biosynthetic pathways, and explorations into the virgin areas of chemotaxonomy and the evolution of phytochemical processes.

General Bibliography and Problems for Students

In order to provide the interested reader with the more extended discussions of natural products, for which this book is a guide, the following list of major works is given. (An asterisk indicates multi-author works; these are inevitably of variable quality.) Discussion of all the topics in this book will be found, as well as references to the original papers, either in these major reviews or in the important multilingual series *Fortschritte der Chemie Organischer Naturstoffe* (consult index, volume 20), edited by L. Zechmeister. The *Quarterly Reviews* of the Chemical Society (London) also often contains reviews on natural products.

(a) BIOGENESIS AND COMPENDIA OF NATURAL COMPOUNDS

Bernfeld, P. (ed.). *Biogenesis of Natural Compounds.* Pergamon, 1963.*

Karrer, W. *Konstitution und Vorkommen der Organischen Pflanzenstoffe (exclusive Alkaloide).* Birkhauser, 1958.

Miller, Max W. (ed.). *Pfizer Handbook of Microbial Metabolites.* McGraw-Hill, 1961.

Richards, J. H. and Hendrickson, J. B. *Biosynthesis of Steroids, Terpenes, and Acetogenins.* Benjamin, 1964.

Robinson, R. *Structural Relations of Natural Products.* Oxford, 1955. (Mainly of historical interest.)

(b) ACETOGENINS

Dean, F. M. *Naturally Occurring Oxygen Ring Compounds.* Butterworth, 1963.

Geissman, T. A. *Chemistry of Flavonoid Compounds.* Macmillan, 1962. *

Ollis, W. D. (ed.). *Recent Developments in the Chemistry of Natural Phenolic Compounds.* Pergamon, 1961.*

Thomson, R. H. *Naturally Occurring Quinones.* Academic Press, 1957.

(c) TERPENES [1]

Simonsen, J. L., and others. *The Terpenes.* 5 vols. Cambridge, 1947–1957.

Fieser, L. F. and M. *Steroids.* Reinhold, 1959.

Karrer, P. and Jucker, E. *Carotenoids.* Rev. ed. by E. A. Braude. American Elsevier, 1950.

Ourisson, G., and others. *Tetracyclic Triterpenes.* Holden-Day, 1964.

Rodd, E. H. (ed.). *Chemistry of Carbon Compounds.* Vol. 2, Part B. American Elsevier, 1953.

(d) ALKALOIDS [2]

Boit, H. G. *Ergebnisse der Alkaloid-Chemie bis 1960.* Akad. Verlag, 1961.

Ginsburg. D. *The Opium Alkaloids.* Interscience, 1962.

Manske, R. H. E., and Holmes, H. L. (eds.). *The Alkaloids: Chemistry and Physiology.* 7 vols. Academic Press, 1950–1960.*

---

[1] Herz, W., *et al.*, The Structure of Tenulin (p. 115): *J. Am. Chem. Soc.*, **84**, 3857 (1962).

[2] Gorman, M., Neuss, N. and Biemann, K. The Structure of Vindoline (p. 164): *J. Am. Chem. Soc.*, **84**, 1058 (1962).

(e) MISCELLANEOUS WORKS

The first two volumes listed in this section are general collections published as commemorative volumes and containing many interesting articles on natural products. Those that follow are a rather arbitrary selection of works devoted to structure elucidation which emphasize natural products problems. (A concise introduction to the main physical techniques employed in discussion in this book can be found in Silverstein, R. M. and Bassler, G. C. *Spectrometric Identification of Organic Compounds*, J. Wiley, 1963.)

Prof. Dr. Arthur Stoll. *Festschrift*. Birkhauser, 1957* (multilingual).
Todd, A. R. (ed.). *Perspectives in Organic Chemistry*. Interscience, 1956.*
Scott, A. I. *Interpretation of the Ultraviolet Spectra of Natural Products*. Pergamon, 1963.
Budzikiewicz, H., and others. *Structure Elucidation of Natural Products by Mass Spectrometry*. 2 vols. Holden-Day, 1964.
Djerassi, C. *Optical Rotary Dispersion*. McGraw-Hill, 1960.
Weissberger, A. (ed.). *Technique of Organic Chemistry*. 11 vols. *Elucidation of Structures by Physical and Chemical Methods*. Vol. 11, 2 Parts. Interscience, 1963.*

Throughout the present text will be found a number of problems of structural logic, mechanistic interpretation, and so forth, which may serve as student exercises. For convenience, these problems are listed here by chapter. An asterisk indicates problems for which the solution is also presented. In the text, a stop-mark (■) will be found at the end of the presentation of evidence, and before the solution, to assist readers interested in solving problems for themselves. Problems are not arranged according to difficulty but only as they arrive in the text. In the literature of natural products it is very uncommon to find cases, useful as problems, in which adequate experimental evidence for a solution is given without structural interpretation. One such case, however, will be found in Raistrick's paper on fuscin, *Biochem. J.*, **48**, 67 (1951); the solution is given in *J. Chem. Soc.*, **1956**, 1028.

CHAPTER 2—BIOGENESIS

1. Dissection of biogenesis of acetogenin structures, Chart 5 and Chapter 3.
2. Dissection of terpene structures into isoprene units, Chart 6 and Chapter 4.
3. Biogenesis of stemmadenine, Chart 10c.
4. Biogenesis of ajmaline, Chart 10c.

2

BIOGENESIS OF NATURAL

PRODUCTS ∎

OUR PRESENT UNDERSTANDING of biochemistry allows a remarkably detailed view of the substances and reactions which make up the chemical machinery of life, the metabolic pathways which utilize food, convert energy, manufacture tissue, and eliminate waste. In the course of this primary chemical activity, however, the metabolic laboratory also creates a number of secondary substances, the biological purpose of which for the most part is obscure. These compounds very commonly possess wonderfully complex structures, however, and thus have long been a source of challenge and stimulation to the organic chemist.

The three major groups of these compounds—the acetogenins, terpenes, and alkaloids—form the substance of the remainder of this book, and their origins in primary metabolism are shown in Chart 1. The chart is arranged to emphasize the basic pool of reactions which produce the major polymeric tissue materials listed at the right and the secondary metabolites, listed below, which are commonly the natural products of the organic chemist, and are the subject of this book.[1]

CHART 1 ■

General Areas of Metabolism in Living Organisms

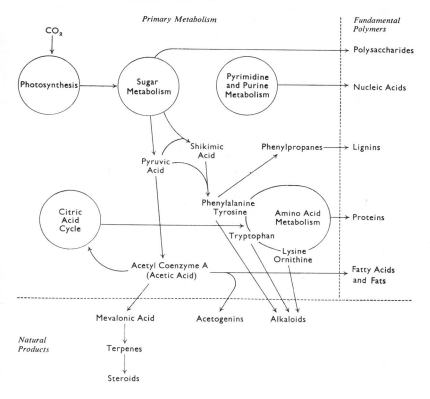

The biogenesis of the natural products is now understood at least in broad outline and often in considerable detail. Simplicity is the most conspicuous feature of their biogenesis, the prodigious diversity of structure among these compounds arising from the interplay of

[1] A few compounds essential to primary metabolism, notably hormones, vitamins, and coenzymes are included as they fall naturally into the mainstream of the discussion and have, independently, chemical interest as well as metabolic importance.

relatively few fundamental reactions and starting materials. Indeed, if these natural products are of only secondary importance to the organism, it is no surprise that a minimum of biochemical elaboration has been expended on their formation.

From the hundreds of compounds identified as the components of primary metabolism only a handful serve as source material for the elaboration of the thousands of known natural products. By far the most important is acetic acid; the other major springboards are the aromatic amino acids trytophan, phenylalanine, and tyrosine, and the aliphatic amino acids, ornithine and lysine. Add to this methionine as a methylating agent and the list is virtually complete.

Most, but possibly not all, of the reactions generating these compounds are enzyme catalyzed, but it must be remembered that the function of an enzyme is only to catalyze, so that the reactions involved are therefore always mechanistically (and stereochemically) reasonable ones in terms of ordinary organic reaction theory and may, for our purposes here, be most conveniently discussed in such terms. That some of the reactions involved in natural product production may not be enzyme catalyzed is suggested, although not proved, by the ability of certain presumed intermediates to react spontaneously under conditions extant in the living cell; if this is true, such natural products could be only metabolic accidents.

The organic chemist has long been intrigued by the synthetic mechanisms used in living systems to produce the many complex organic molecules found in nature. The first forays into biogenetic speculation began with the recognition of common structural features among the compounds produced by closely related natural organisms. This recognition led to inferences of a relatively simple common origin for these compounds in ordinary biochemical starting materials reacting via a sequence of transformations regarded as feasible in the living cell.

These speculations have had a long and lively history and have led to the successful construction of some remarkably simple laboratory syntheses of complex natural molecules, modeled on biogenetic lines, as well as furnishing a valuable line of hypothesis to direct the more recent experimental efforts in delineating *in vivo* paths of biosynthesis. Most fruitful among these efforts have been feeding experiments with radioactive precursors followed by isolation of the radioactive natural compound and chemical degradation to isolate particular atoms and examine their radioactivity. It is a tribute to the authors of these biogenetic schemes that they have so often been proved correct.

2–1 The Reactions of Biosynthesis: A Survey

If the number of starting materials for natural product biosynthesis is severely limited, so too is the variety of chemical reactions utilized en route. In order to emphasize this simplicity as well as to simplify the subsequent discussion of detailed biosynthesis of the major families, it will be useful here to summarize the basic chemical processes leading to their characteristic carbon skeletons.

(1) ACETOGENINS   Fundamentally, these compounds are formed by head-to-tail condensations of acetic acid into a linear chain of alternating ketone and methylene groups, viz.,

$$CH_3COCH_2COCH_2COCH_2 \ldots COOH.$$

Such chains then cyclize via an aldol condensation between an enolic methylene and a ketone six atoms away.

(2) TERPENES   Terpenoid structures are characterized by chains formed from the head-to-tail union of "isoprene units,"

$$
\begin{array}{c}
C \\
| \\
C-C-C-C
\end{array}
$$

The detailed biosynthesis of this five-carbon basic unit involves not a linear but a branched condensation of three acetic acid molecules to

$$
\begin{array}{c}
CH_2-COOH \\
| \\
CH_3-C-CH_2-COOH \\
| \\
OH
\end{array}
$$

followed by decarboxylation-dehydration. The chains of from two to eight isoprene units then formed may subsequently cyclize also.

(3) ALKALOIDS   The basic operation here simply involves the conversion of amino acids, $RCH(NH_2)COOH$, to amines, RCH_2NH_2, and aldehydes, $RCHO$, and uniting the two via a Mannich reaction.

In each of these families a very few secondary reactions are then found which serve to elaborate the individual differences between the members. These reactions are illustrated in Chart 2 and summarized as follows:

(1) ALKYLATION   Virtually all the biosynthetic alkylations

CHART 2 ■

Common Secondary Biosynthetic Reactions

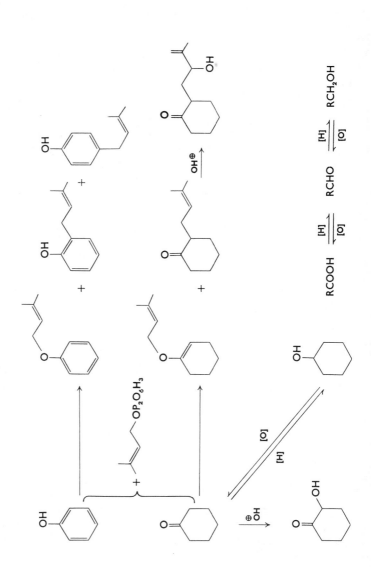

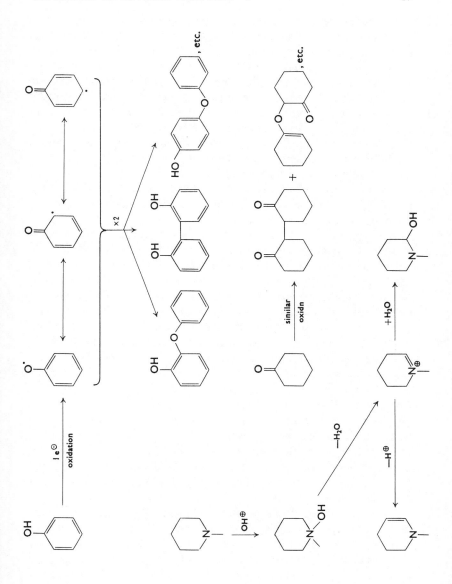

encountered are methylations or isopentenylations and can occur on amines or at either the carbon or oxygen of an enolate (or phenolate) anion, although methylation is practically the only alkylation of amines. Biochemically, methylation involves the transfer of a methyl group from the sulfur of methionine while isopentenylation occurs via attack of an enolate (or sometimes an olefin) on isopentenyl pyrophosphate, the fundamental terpene precursor discussed in more detail in the section on terpene biosynthesis (p. 33).

(2) REDUCTION   Reduction of carbonyl groups to alcohols (including the reduction of carboxyl) is a common process and so is the mechanistically analogous reduction of double bonds conjugated to carbonyl groups. The reverse reactions, the corresponding oxidations, are also common, most biochemical transformations being reversible reactions.

(3) OXIDATION   It is convenient to distinguish between one-electron and two-electron oxidations here, although it is not known in particular cases which type obtains, since either can be envisioned mechanistically to account for any given oxidation product. The two-electron oxidations are most easily grouped as attacks of an equivalent of $HO^\oplus$ on an amine, an enolate anion, or an olefin. The latter two cases lead to simple hydroxylation; the former leads first to an N-oxide and then to an imine, eneamine, or carbinolamine, as shown in Chart 2. One-electron oxidation of an enolate (or phenolate) anion leads to the enolic radical which may then couple with another of its kind, forming a dimeric molecule the halves of which are joined by C—O or C—C bonds (see Chart 2).

(4) SPONTANEOUS REACTIONS   A number of reactions, known to be facile under relatively neutral conditions at room temperature, must certainly be allowable as biosynthetic steps; such reactions include aldol condensations, dehydration of alcohols β- to carbonyl groups, decarboxylation of β-keto acids and spontaneous closure of γ- and δ-lactones and lactams.

In the remainder of this chapter the biogenesis of each of the three major families is separately considered in more detail, the operation of variants delineated, and the application of the general secondary reactions considered. Next is a survey of the major structural types produced within each family, providing a concise scan of the scope of natural product molecules. The phenylpropane (ϕ-C_3) compounds considered in Section 2–2 are more nearly a part of primary metabolism, closely allied to lignin production. Because they are sometimes themselves involved in the biosynthesis of other natural products they are examined ahead of the three main families.

2-2   Origin of Aromatic Rings; The Phenylpropanes

The majority of natural products incorporate at least one benzene ring. Thus it is reasonable to inquire first into the origin of such rings in nature. Because the substantial resonance energy in the benzene ring would lead one to anticipate, a priori, a wide variety of originating modes from nonaromatic substances, such as sugars, it comes as an initial surprise to find that there are only two routes accounting for almost every natural aromatic carbocycle. One of these is fundamental to acetogenin biosynthesis and is discussed on pages 24–25; the other is the central mode in primary metabolism and is outlined in Chart 3.

This route, originally delineated by Davis and Sprinson and often called the shikimic acid pathway, involves an initial aldol condensation of "phospho-enol pyruvate" [2] with a tetrose [1] to yield a seven-carbon sugar [3]. This undergoes an aldol cyclization to dehydroquinic acid [4], which is transformed first into shikimic acid [5] and then into prephenic acid [6] by further addition of the pyruvate enol [2]. The subsequent aromatization of prephenic acid yields phenylpyruvic acid [7] from which phenylalanine [8] is formed by reductive transamination; further deamination yields cinnamic acid [9], and a parallel series occurs at the higher oxidation state of tyrosine [10].

The benzene rings produced in this way retain at most the 3,4,5-trihydroxy oxidation pattern of shikimic acid; otherwise this pattern is truncated to 3,4-dihydroxy, 4-hydroxy, or no oxidation. Many simple natural compounds are derived from this route bearing the phenyl-propane (ϕ-C_3) skeleton and these particular patterns of oxidation. Their variety arises from the operation of the several biogenetic reactions previously discussed and may be discerned in the structures of *p*-coumaric [11] and ferulic [12] acids and syringyl alcohol [14]. The fungus terphenyls (cf. [15]) are readily derived by aldol dimerization of the parent phenylpyruvic acids. Oxidative coupling of these phenols can occur at the β-position of the C_3 side chain, as from [16], and affords a number of dimers known as lignanes, of which sesamin [17] is representative. The major biochemical purpose for these oxidative couplings of phenylpropane precursors is probably the

CHART 3 ■

The Shikimic Acid Pathway to Aromatic Compounds

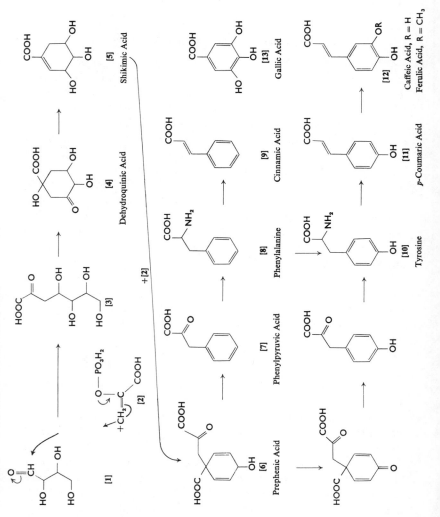

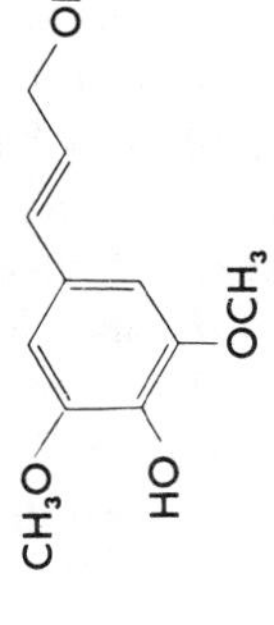

[16]

[14]
Syringyl Alcohol

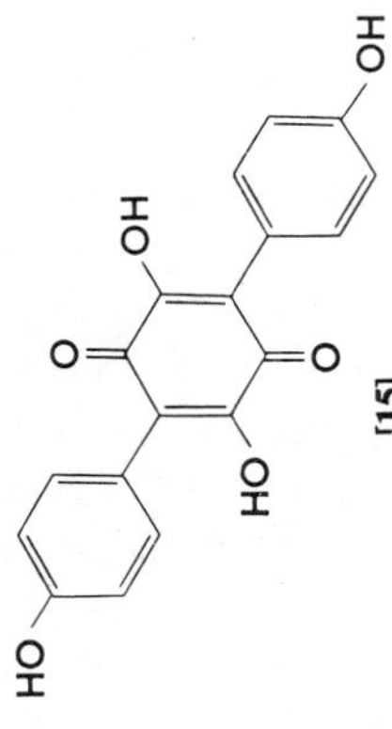

[15]
Atromentin

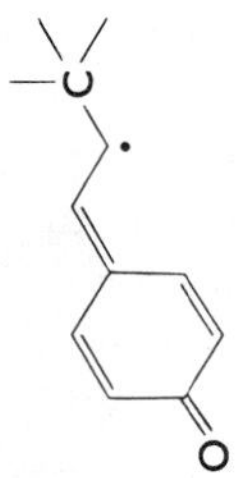

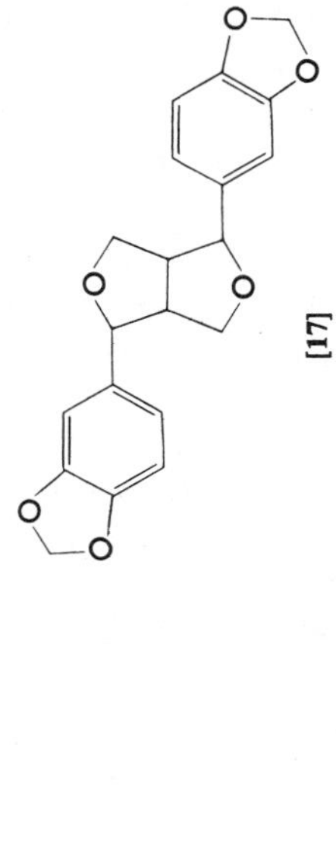

[17]
Sesamin

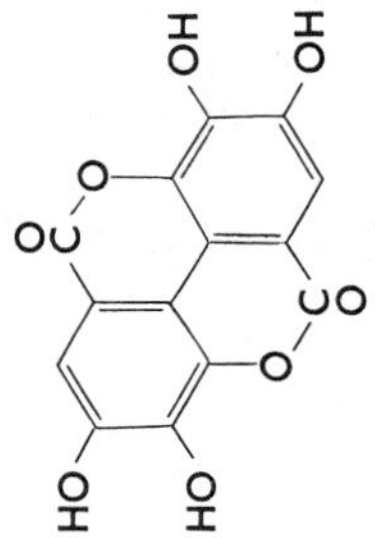

[18]
Ellagic Acid

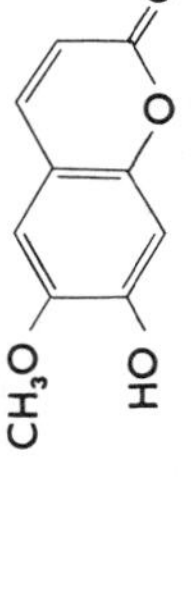

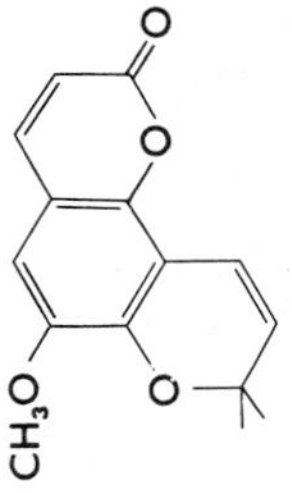

[22]
Braylin

[21]
Bergamotin

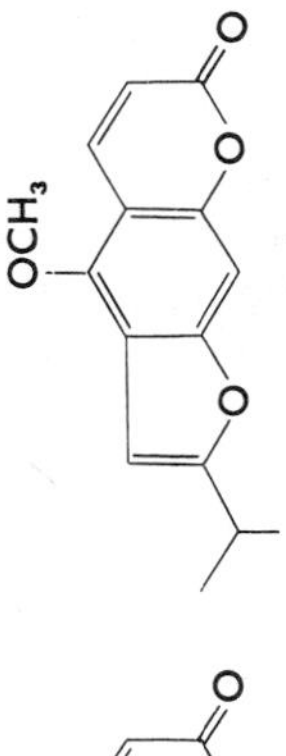

[20]
Xanthoxyletin

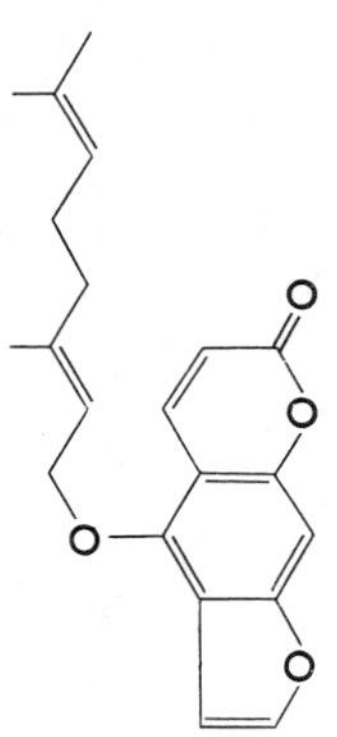

[19]
Scopoletin

production of the lignin polymers which are the fundamental basis of structural tissue in all plants.

Oxidative coupling of such phenols is the key to lignin formation and to that of the tannins as well, which are largely derived from gallic acid [13], itself a shikimic acid product. Studies of tannins have produced a series of natural compounds, such as ellagic acid [18], which are clearly gallic acid oxidative coupling products.

The phenylpropane skeleton is also probably the basis of most of the nearly one hundred known natural coumarins, which are apparently produced by oxidative cyclization of a cinnamic acid. The coumarins display an impressive selection of variants derived from alkylation as compounds [19]–[22] will testify. The methyl and isopentenyl substituents are always found at aromatic sites consonant with the mechanism of enol alkylation.

2–3   Biosynthesis of the Acetogenins

The fatty acids are characterized by a long hydrocarbon chain, without branching, culminating in a carboxyl group. As a group of compounds with an important metabolic role they have long been studied by biochemists with the result that their biosynthesis is now well understood and is summarized in Chart 4. The basic skeletal construction involves carbon–carbon bond formation via an aldol-type condensation. These condensations are utilized to add sequentially the two-carbon units of acetic acid head-to-tail into a long linear chain.

Activation of acetic acid to afford sufficient reactivity for an aldol condensation is effected in two ways: first, the carbonyl is activated for nucleophilic attack by conversion of the carboxyl to a thiol ester; second, the adjacent methylene is further activated by carbonation to a malonic acid thiol ester. Following the basic carbon–carbon bond formation are a series of reductions transforming the initial β-keto-thiol ester product to a saturated thiol ester in preparation for subsubsequent chain-lengthening in an analogous manner. The entire series of reactions is believed to occur while the growing acid is esterified to the –SH group of an enzyme.

The acetogenins are a large and varied group (over a thousand) of natural compounds which owe their formation to a few reactions of a long linear polyacetyl chain,

$$CH_3—CO—(CH_2—CO)_n—CH_2—COOH,$$

CHART 4 ■

Biosynthesis of Fatty Acids and Acetogenins

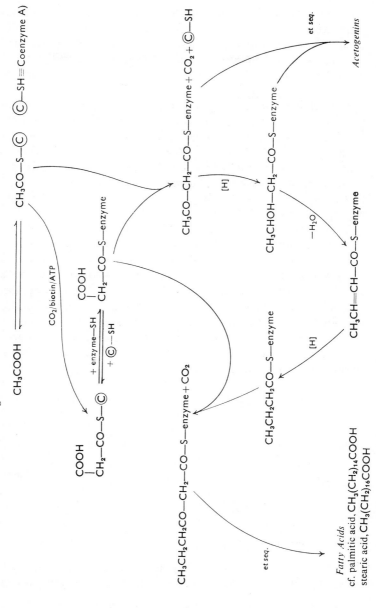

CH_3COOH

$CH_3CO—S—Ⓒ$ $Ⓒ—SH \equiv Coenzyme\ A)$

$CO_2/biotin/ATP$

$+\ enzyme—SH$
$+\ Ⓒ—SH$

COOH
|
$CH_2—CO—S—Ⓒ$

COOH
|
$CH_2—CO—S—enzyme$

$CH_3CO—CH_2—CO—S—enzyme + CO_2 + Ⓒ—SH$

[H]

$CH_3CHOH—CH_2—CO—S—enzyme$

$-H_2O$

$CH_3CH=CH—CO—S—enzyme$

[H]

$CH_3CH_2CH_2CO—S—enzyme$

$CH_3CH_2CH_2CO—CH_2—CO—S—enzyme + CO_2$

et seq.

Acetogenins

et seq.

Fatty Acids
cf. palmitic acid, $CH_3(CH_2)_{14}COOH$
stearic acid, $CH_3(CH_2)_{16}COOH$

which can be seen to arise via the same head-to-tail condensation sequence of acetate units without the reductive steps intervening in each cycle. The subsequent elaboration of such poly-β-keto-methylene chains, first delineated by Birch, occurs by the operation of the several general reactions outlined above[2] and may conveniently be summarized as follows:

1. Each methylene in the chain is activated by adjacent carbonyls and may react with alkylating agents or $^{\oplus}OH$, giving rise to substitution at the methylene sites by methyl, isopentenyl, or hydroxyl. (Substitution by methylene or one-carbon fragments of higher oxidation state is occasionally found, although biosynthetically this may well represent only methylation with later oxidation.)

2. Two chains may be coupled, again at methylene sites, by the oxidative coupling reaction of enols.

3. A keto group occasionally will be found reduced to an alcohol and/or subsequently dehydrated (being a β-hydroxy ketone).

4. The terminal carboxyl may be lost as CO_2 since the compounds are β-keto-acids. Hence, whereas chains normally contain an even number of carbons, decarboxylated compounds are odd numbered.

5. Normally the chain is initiated by acetyl coenzyme A and extended by malonyl units, but other acids may initiate the chain. However, even these are virtually limited to propionic acid and the ϕ-C_3 acids of shikimic origin, particularly the cinnamic and phenyl pyruvic acids of Chart 3.

$$\overset{8}{C}H_3-\overset{7}{C}O-\overset{6}{C}H_2-\overset{5}{C}O-\overset{4}{C}H_2-\overset{3}{C}O-\overset{2}{C}H_2-\overset{1}{C}O-SR$$

[23]

1–6   2–7

[24]
Phloracetophenone

[25]
Orsellinic Acid

6. The major source of variety lies in internal cyclization of the polyacetyl chain in which one active (enolic) methylene attacks a

---

[2] It is probable that most of these reactions occur while the polyacetyl chain is still attached to the enzyme, for such polyketones are very unstable *in vitro*.

carbonyl six atoms away in an intramolecular aldol condensation—usually leading to a benzene ring—although attack by the enolic oxygen often occurs, forming pyran rings instead. These cyclizations form the second fundamental mode in nature for the synthesis of benzene rings. In fact, the polyacetyl cyclization and the shikimic acid pathway appear to be the only routes to aromatics in living organisms. Two kinds of polyacetyl cyclization to benzene rings are shown with the tetraacetyl chain [23] and yield two families of acetogenins of which the natural products phloracetophenone [24] and orsellinic acid [25] are the simplest representatives.

The six preceding "rules" produce nearly 90 per cent of all acetogenins, and their biogenesis may be summarized in the convenient notational convention [26]. It is now possible to dissect the structures of acetogenins in order to see the operation of these several biogenetic

[26]

variants, for their structures should reveal the folded but linear chain with oxygen "markers" at the ketone carbons of the original chain (except where cyclizations remove them), and methyl or isoprenyl branches or coupling dimerization at the original methylene sites.

A survey of the major structural groups of acetogenins is shown in the collection of illustrative examples in Chart 5[3] for some of which the biosynthetic dissection (cf. [31]) is indicated; for the others it should be clear from the formulas. The chemistry of these groups is the substance of Chapter 3, but a brief delineation of the main structural families here may serve to clarify and organize this multifarious collection.

The simplest acetogenins (biogenetically!) after the fatty acids are the linear polyacetylenes [27]–[28] and the antibiotic macrolides [29] (as well as shorter chains initiated by aromatic acids of shikimic origin such as the Polynesian Kawa root ingredient [30]). Chain-extension and alkylation of the simpler acetyl-phloroglucinols (cf. [24]) creates the chromones [31] and the constituents of ferns [32] and hops [33],

[3] The formulas of acetogenins in this book are always shown with the chain-initiating group (methyl or phenyl) at the left to display their biogenetic unity.

CHART 5 (PART I) ■

Representative Examples of the Major Structural Types of Acetogenins

Polyacetylenes:

$CH_3-(C\equiv C)_3-CH=CH-CH_2OH$

[27]

Matricarianol

$R-C\equiv C-C\equiv C-CH=C-CH=CH-CH-CH_2-CH_2-COOH$
$\qquad\qquad\qquad\qquad\qquad\qquad\quad |\!-OH$

[28]

Nemotinic Acid, R = H
Odyssic Acid, R = CH$_3$

Macrolides:

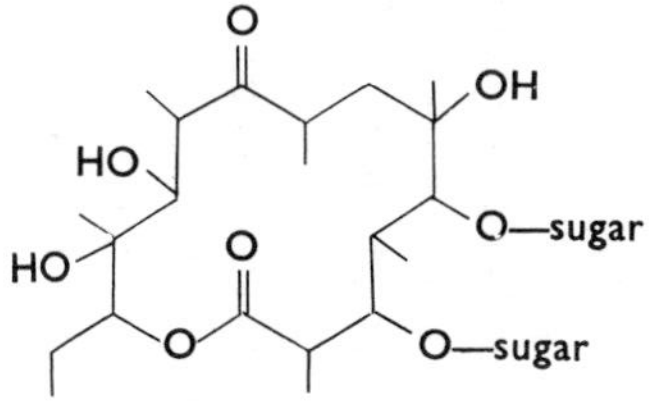

[29]

Erythromycin

Chromones:

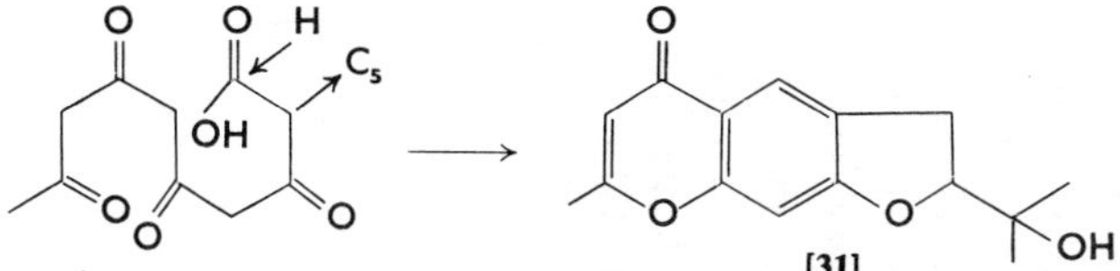

[31]

Visnamminol

ϕ-C_n:

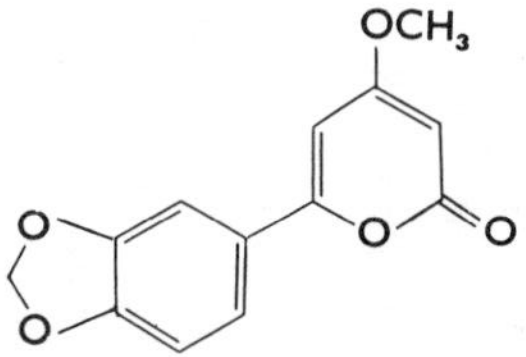

[30]

Methoxy-paracotoin

Acyl-phloroglucinols:

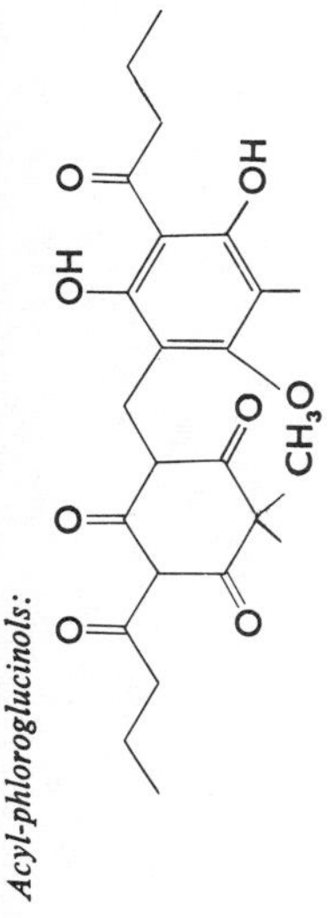

[32]
Flavaspidic Acid

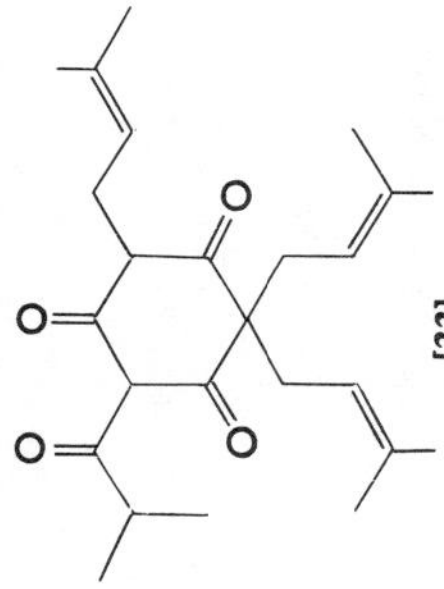

[33]
Colupulone

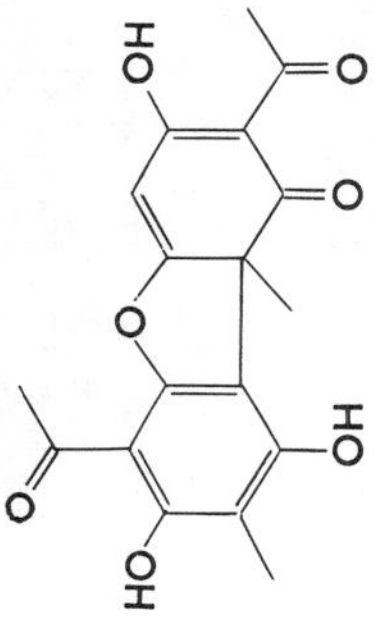

[34]
Usnic Acid

Benzophenones:

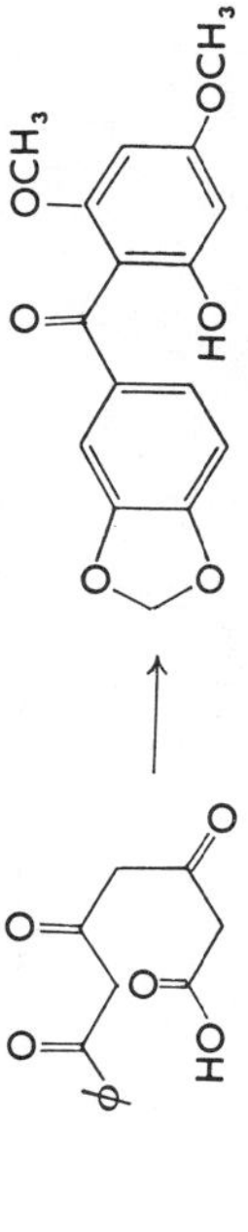

[35]
Protocotoin

Xanthones:

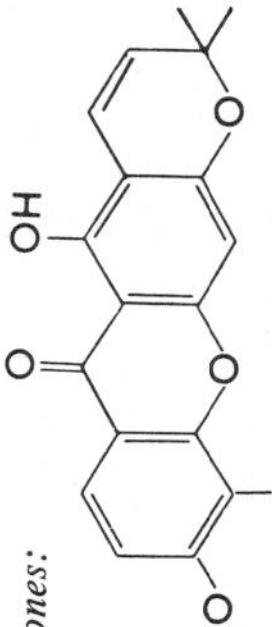

[36]
Jacareubin

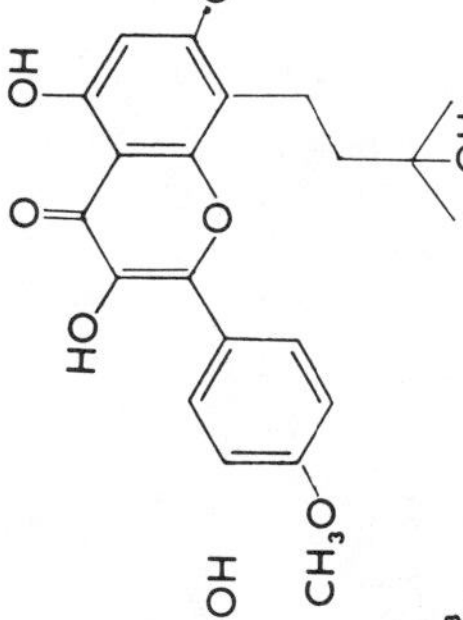

[38]
Icaritin

[39]
Matteucinol

Flavanoids:

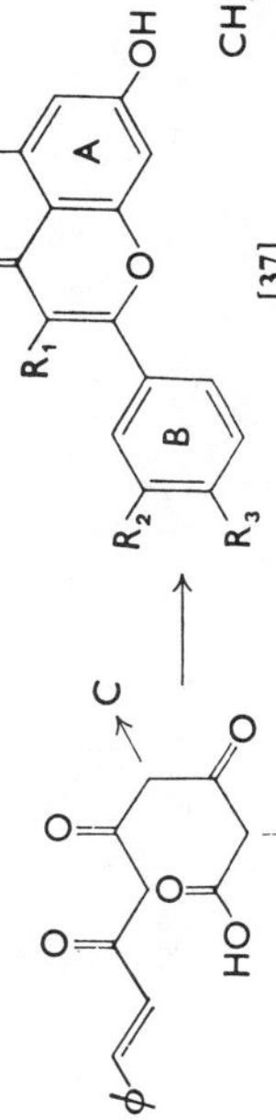

[37]

	R_1	R_2	R_3
Chrysin	H	H	H
Apigenin	H	H	OH
Luteolin	H	OH	OH
Quercetin	OH	OH	OH

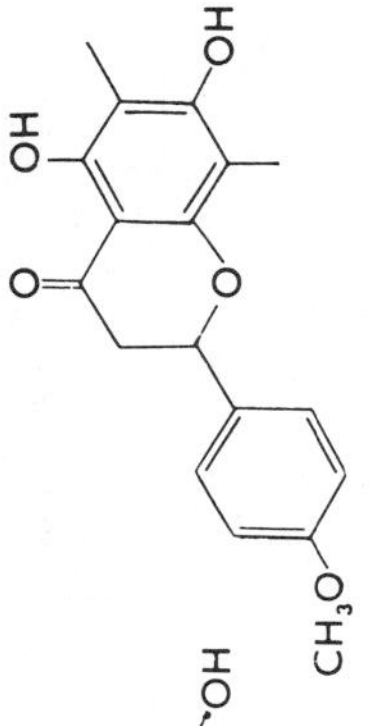

CHART 5 (PART II) ■

Isoflavonoids:

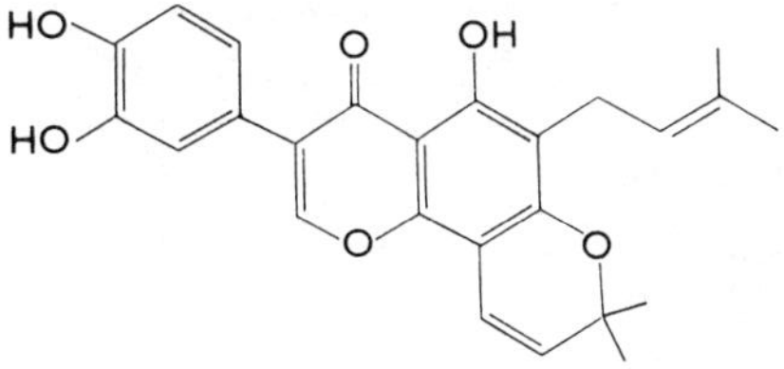

[40]
Pomiferin

Rotenoids:

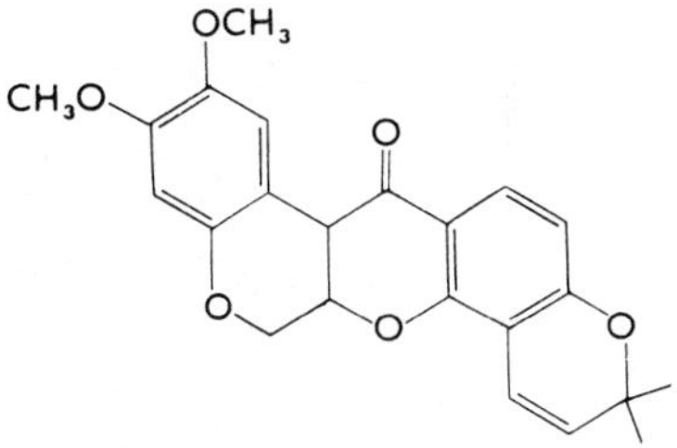

[41]
Deguelin

Salicylic Acids and Catechols:

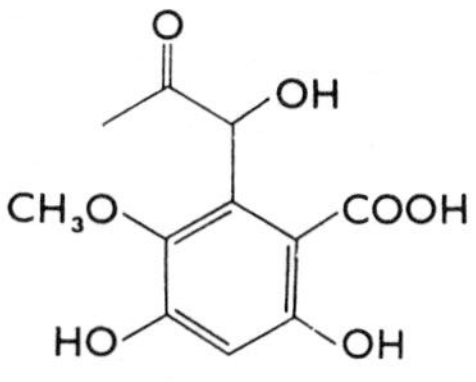

[42]
Ustic Acid

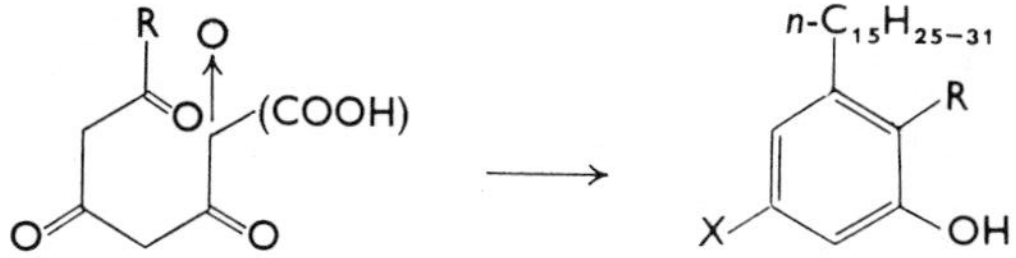

[43]

Anacardic Acid: $R = COOH; x = H$
Cardol: $R = H; x = OH$
Urushiol: $R = OH; x = H$

Depsides:

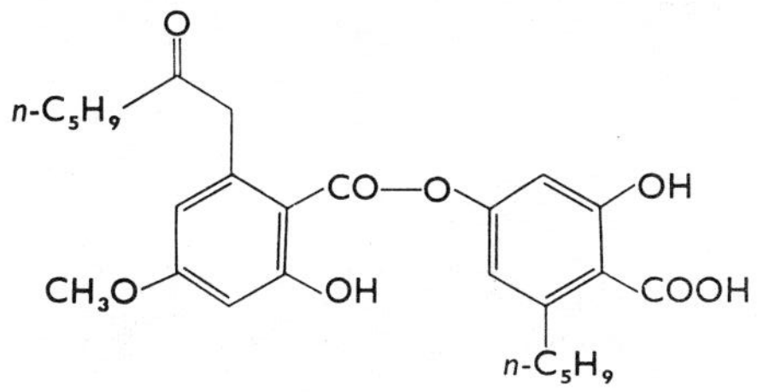

[44]
Microphyllic Acid

Depsidones (and related dibenzofurans):

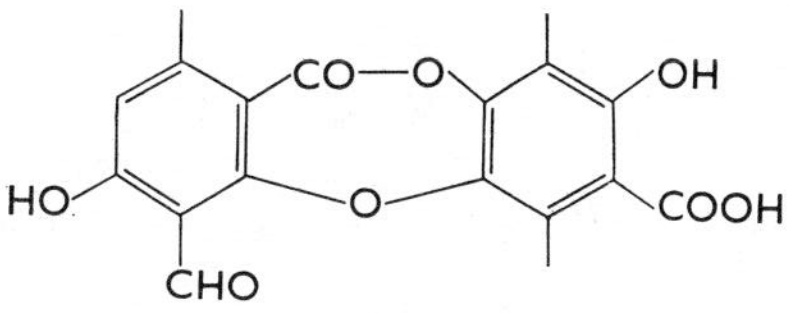

[45]
Virensic Acid

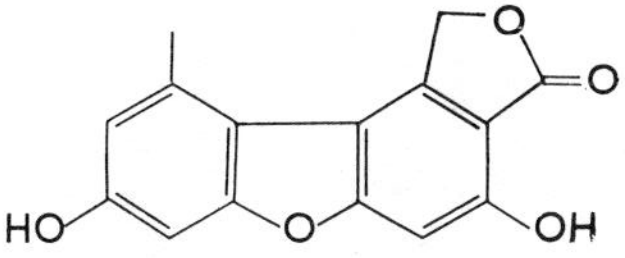

[46]
Strepsilin

Stilbenes:

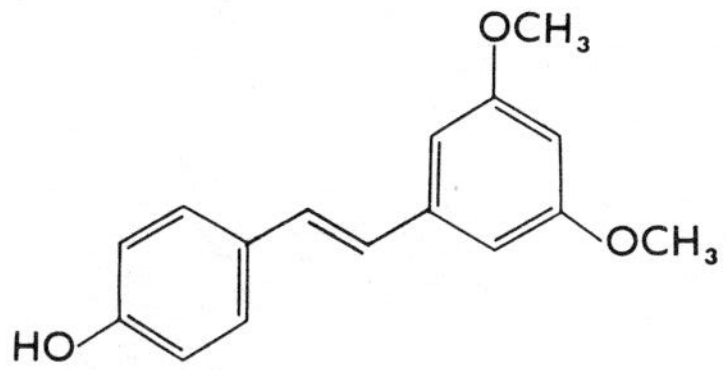

[47]
Pterostilbene

Naphtholenes and Naphthoquinones:

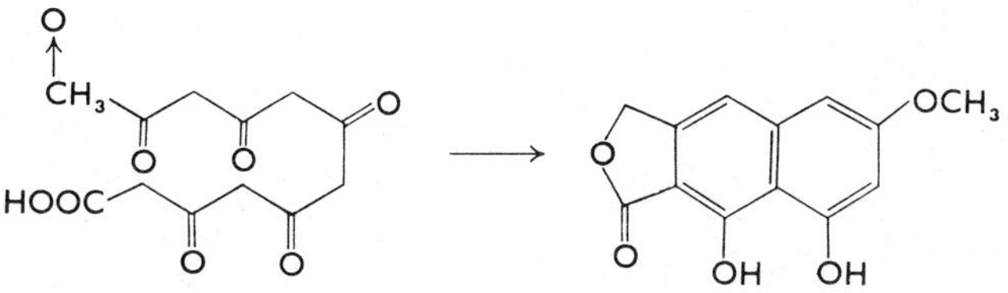

[48]
α-Sorigenin

CHART 5 (PART III) ■

Anthraquinones:

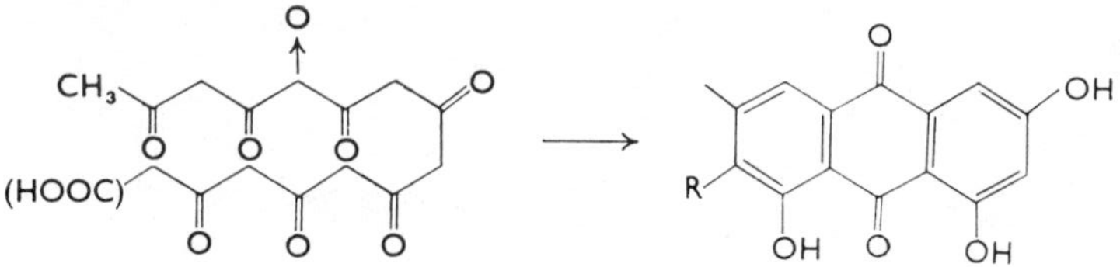

[49]
Emodin, R = H
Endocrocin, R = COOH

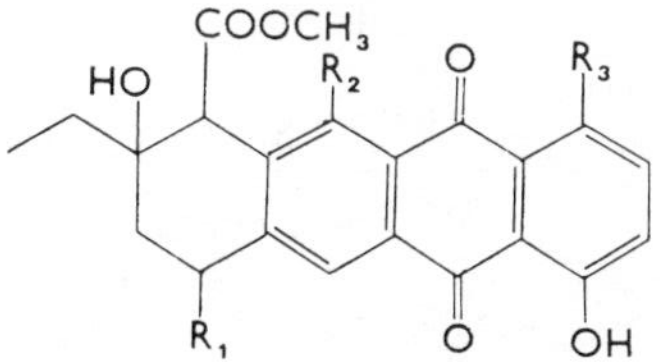

Mycinones: R's = H, OH

[50]

Tetracyclines:

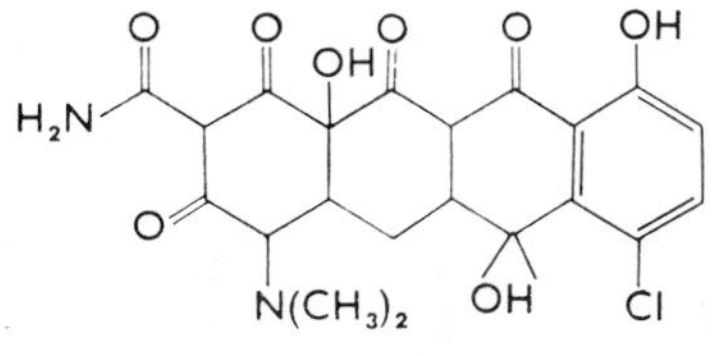

[52]
Aureomycin

Dimeric Quinones:

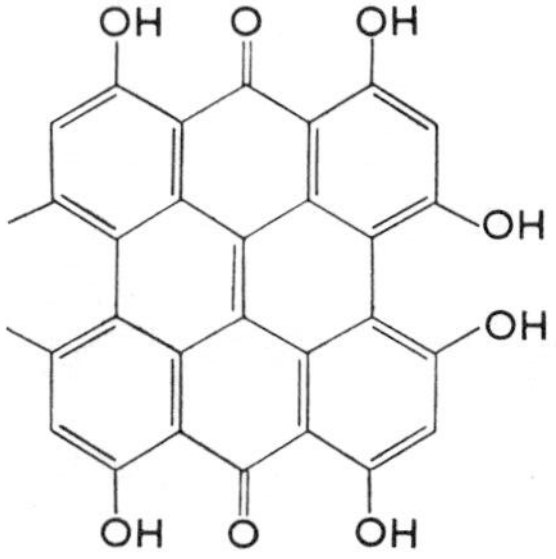

[51]
Hypericin.

[54]
Fulvic Acid

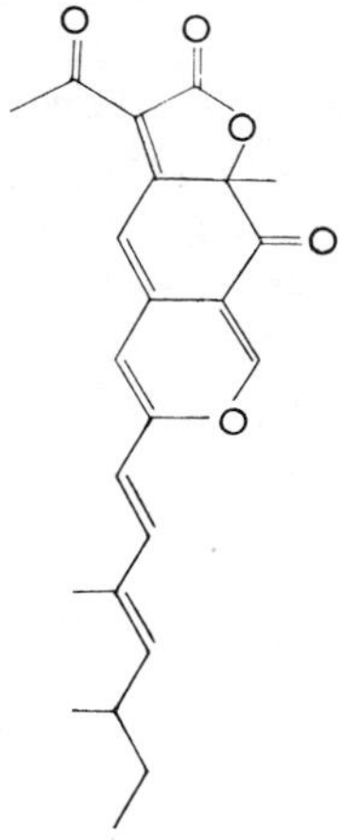

[56]
Rotiorin

[53]
Alternariol

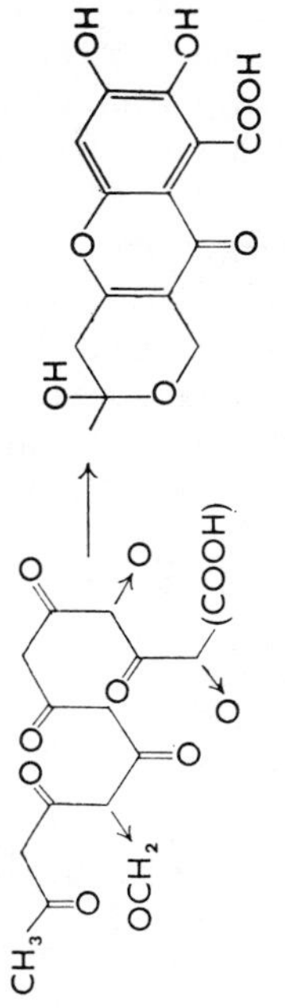

[55]
Atrovenetin

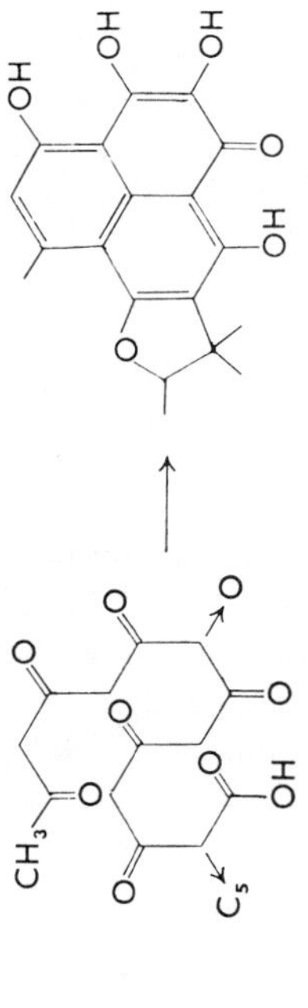

Polycyclic Aromatics:

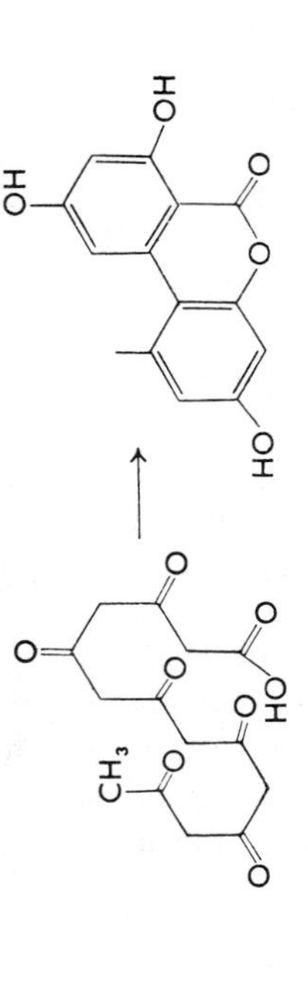

whereas oxidative dimerization yields the widespread lichen product, usnic acid [34]. The dozen or so natural benzophenones [35] arise from chain initiation by a benzoic acid of shikimic origin, and at least some of the (about 15) natural xanthones [36] appear to be intramolecular oxidative coupling products from them.

The numerous (perhaps 300) family of flower and autumn-color pigments known collectively as flavonoids ([37]–[39]) arise from chain extension of shikimic-derived cinnamic acids by three malonate units and differ in the oxidation state of the three (cinnamic) carbons of the pyran ring, as well as combinations of methyl and isopentenyl substitution on the A ring (of acetic origin) and the –OH or –OCH₃ sites

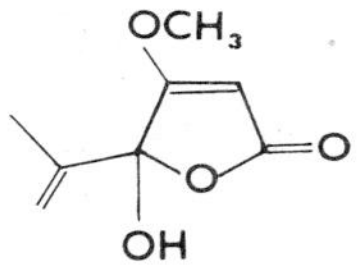

[57]
Penicillic Acid

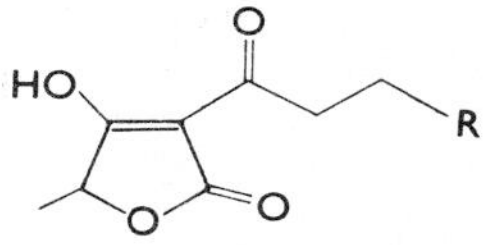

[58]
Carolic Acid, R =CH₂OH
Carolinic Acid, R =COOH
Terrestric Acid, R = CHOHCH₂CH₃

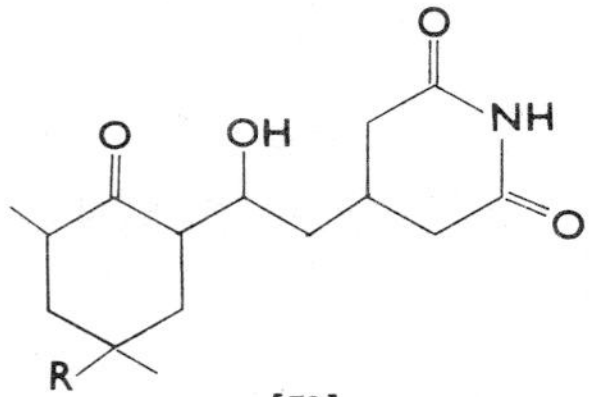

[59]
Actidione, R = H
Streptovitacin. A, R = OH

on the B ring (of shikimic origin). The number of structures possible utilizing the "allowed" variations on the one flavonoid carbon skeleton is obviously immense. The forty-odd isoflavonoids [40] arise by a phenyl migration from the flavonoid skeleton and the rotenoids [41] from these by one-carbon alkylation and cyclization.

On the other hand, cyclization of the polyacetyl chain in the orsellinic acid manner leads to a host of simple fungal products such as [42] and chain initiation by a fatty acid (and subsequent oxidative decarboxylation) affords the poison-ivy vesicants [43]. Lichens are the source of numerous depsides [44] and their oxidative coupling products, depsidones [45]; several dibenzofurans [46] are obviously closely related. Initiation of the polyacetyl chains by cinnamic acids yields the natural stilbenes [47] usually found in the heartwood of pine (often along with biosynthetically analogous flavonoids).

Finally, polyacetyl cyclizations of greater virtuosity lead to a wide assortment ([48]–[56]) of polycyclic aromatics. The anthraquinones [49] are the most numerous of these (over a hundred), turning up as mold and insect pigments frequently, along with the somewhat less varied naphthalenes [48] and naphthoquinones. Oxidative dimers of these [51] are also common. The upper limit in molecular size for acetogenins is about ten linear acetyl units, as exemplified by the mycinones [50] and the tetracycline antibiotics [52].

It is convenient to include here for survey purposes a number of small structure groups of obscure biogenesis which nonetheless appear related to the foregoing acetogenins. The four tropolone acids (cf. [3–82]), pencillic acid [57], and patulin [3–86] are known by C^{14}-tracer work to be acetogenins, as is partly true of the tetronic acids [58]. A number of simple natural cyclopentanes are known (cf. [3–41]) as well as a family of antibiotic imides, the actidones [59], but it is surprising how very few seriously anomalous structures are to be found among natural products.

2–4   Biosynthesis of Terpenes and Steroids

Many of the aliphatic constituents of the essential oils from fragrant plants were early grouped into a family when it was recognized that they all had empirical formulas containing a multiple of five carbons.

[60]
Isoprene

[61]
Rubber

Thus the C_{10} compounds are called monoterpenes; C_{15}, sesquiterpenes; C_{20}, diterpenes; and C_{30}, triterpenes. The monoterpenes were for many years the most extensively studied terpenes and as their structures became known, it became clear that they bore a more remarkable familial resemblance, in that their structures were composed of isopentane skeletal units usually linked together head to tail.

CHART 6 ■

Representative Examples of Terpenes

Monoterpenes:

[62]
Myrcene

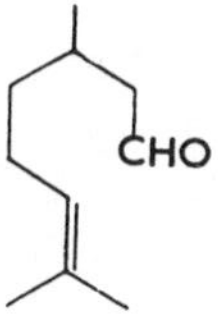

[63]
Citronellal

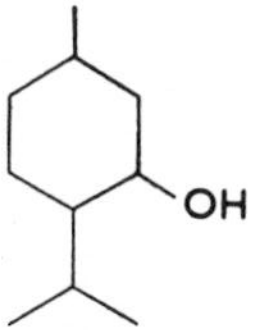

[64]
Menthol

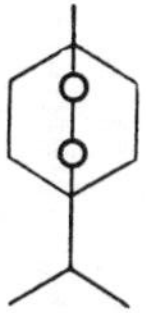

[65]
Ascaridole

[66]
Camphene

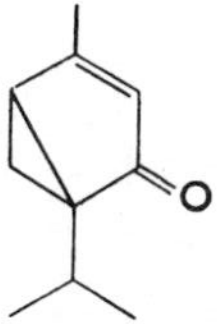

[67]
Umbellulone

Sesquiterpenes:

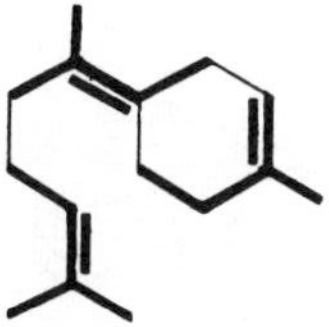

[68]
Bisabolene

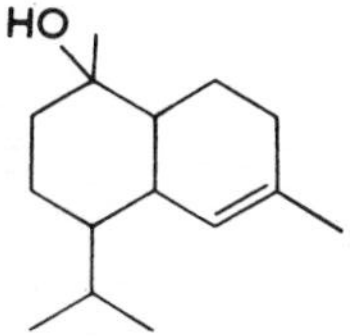

[69]
α-Cadinol

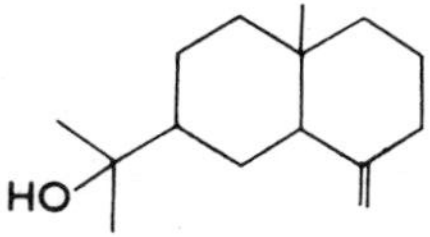

[70]
Eudesmol

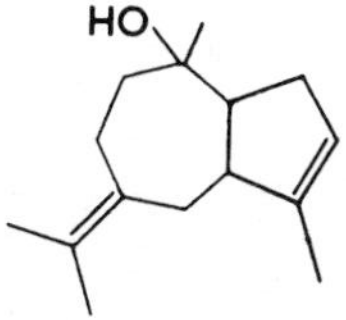

[71]
Partheniol

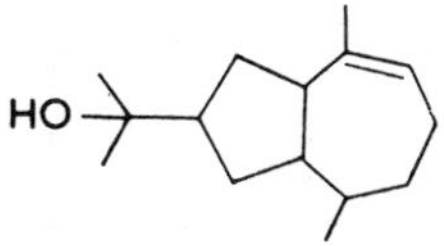

[72]
Hinesol

Diterpenes:

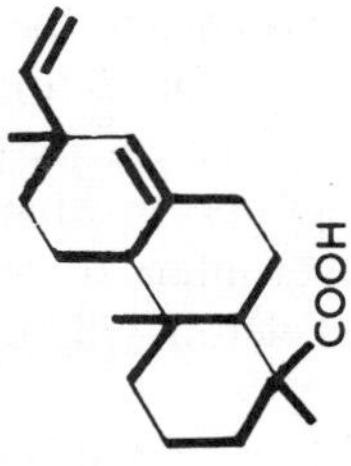

[73]
Dextropimaric Acid

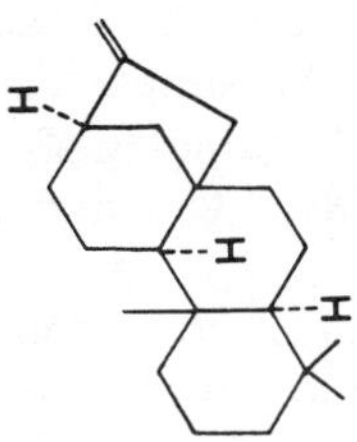

[74]
Phyllocladene

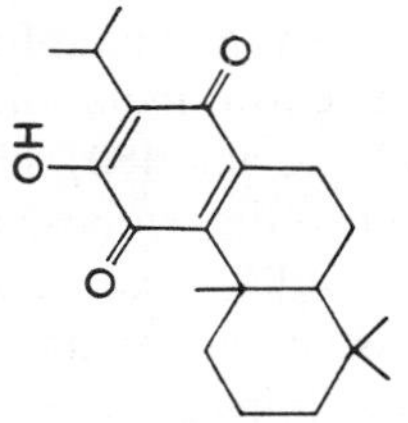

[75]
Royleanone

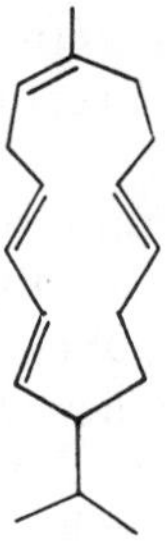

[76]
Cembrene

Triterpenes:

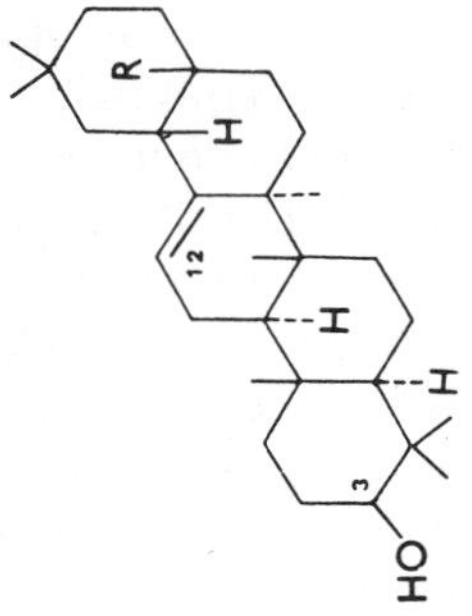

[77]
β-Amyrin, R = CH₃
Oleanolic Acid, R = COOH

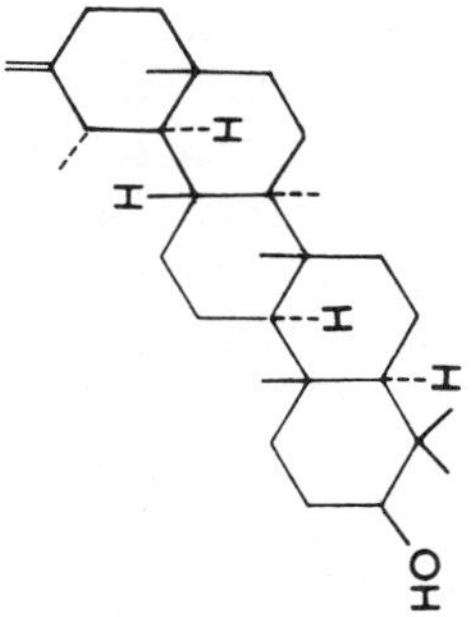

[78]
Taraxasterol

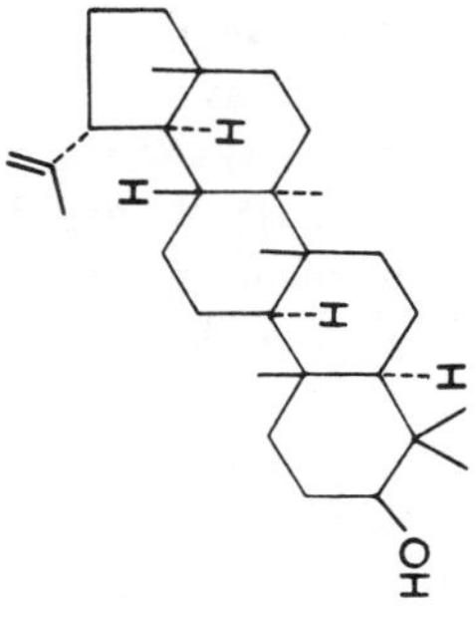

[79]
Lupeol

This ubiquitous and striking feature was formulated as the "isoprene rule" and has served as a potent guide in assigning structures to subsequent and larger terpene molecules under study.

Isoprene itself [60] is formed by distillation of natural rubber, now known to be a linear, head-to-tail polymer [61] of isoprene units. The isoprene units can be discerned in the structures of representative terpenes in Chart 6, and similar dissection (see heavy bonds in examples [62], [68], and [73]) of the structures discussed in Chapter 4 will serve as a demonstration of the generality of the rule as well as an interesting intellectual exercise.

That these repeating isoprene units reflect a unified biogenesis for the terpenes has, of course, been suspected for a long time, but the actual biological isoprene unit and its mode of synthesis and coupling have only recently been elucidated (Chart 7). The central metabolite in terpene biosynthesis is now known to be mevalonic acid [80], first isolated by Folkers' group in 1956. This intermediate is biosynthesized irreversibly from acetate (as outlined in Chart 7). Unlike acetogenin biosynthesis, which it closely resembles mechanistically, mevalonic acid formation proceeds by an aldol condensation yielding a branched chain, and it is this feature which basically distinguishes the terpenes from the linear-chain acetogenins. The subsequent transformation to isopentenyl pyrophosphate [83] yields the true biological isoprene unit in a form activated for the enol alkylations observed in the acetogenins—as well as for successive linking into the simple natural alcohols geraniol (C_{10}), farnesol (C_{15}), and geranylgeraniol (C_{20})—the pyrophosphate esters of which are shown as [84], [85], and [86], respectively.

This attack of the olefinic π-electrons on an electron-deficient carbon, such as that bearing the pyrophosphate-leaving group, is a common biosynthetic reaction in the terpenes. When it occurs internally on these intermediate pyrophosphates it can yield cyclic products as illustrated by the conversion of [84] to limonene [87], or [85] to bisabolene [68], and ultimately to bicyclic sesquiterpenes of the cadinane skeleton [69]. The allylic double bond of [85] must be *cis* for these six-membered cyclizations, but the natural *trans* form in [85] also allows cyclization to the cyclodecane intermediate [89] in which the double bonds are oriented so as to facilitate further internal condensation to sesquiterpenes with the eudesmane skeleton [70] or the guaiane skeletons (cf. [71] or guaiol [4–21]).

While the foregoing cyclization by expulsion of the pyrophosphate is

CHART 7 ■ *Biosynthesis of Terpenes*

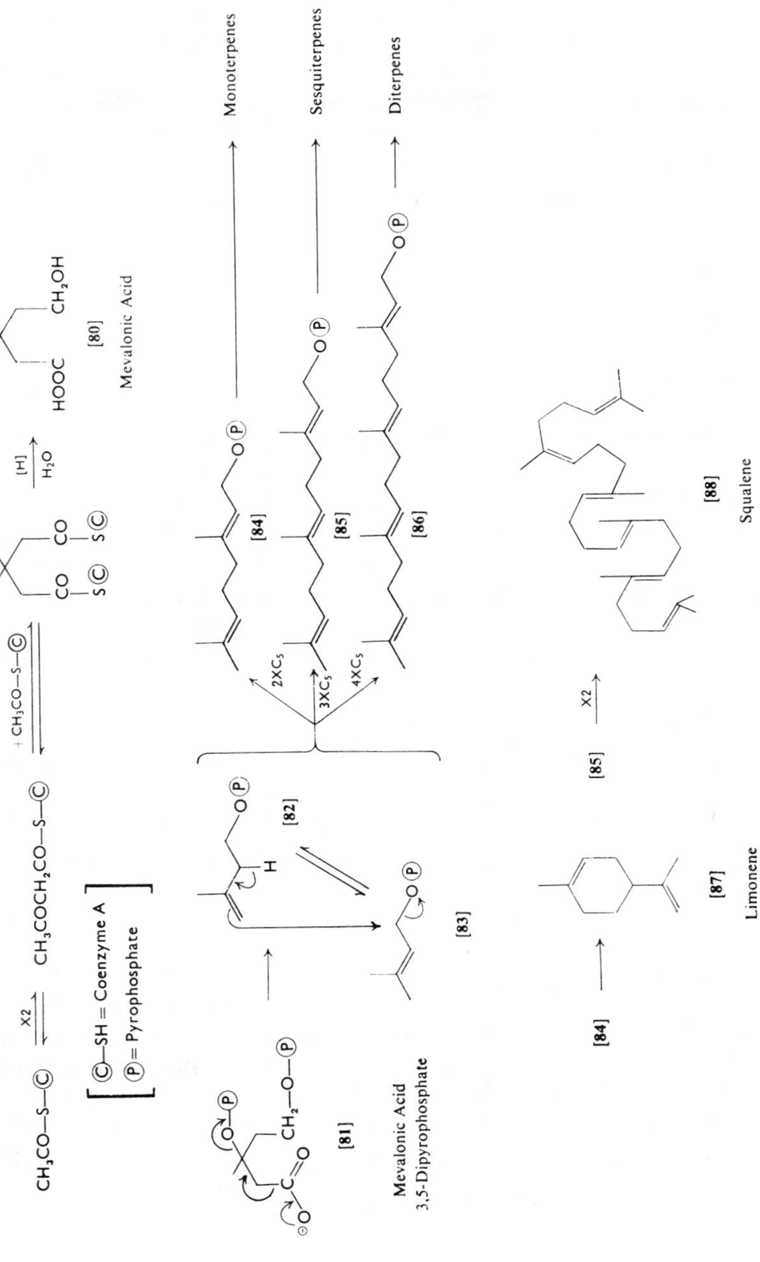
CH3CO—S—Ⓒ
X2
CH3COCH2CO—S—Ⓒ
+ CH3CO—S—Ⓒ
OH
CO
SⒸ
[H]
H2O
HOOC
CH2OH
[80]
Mevalonic Acid
Ⓒ—SH = Coenzyme A
Ⓟ = Pyrophosphate
CH2—O—Ⓟ
[81]
Mevalonic Acid
3,5-Dipyrophosphate
[82]
[83]
2XC5
3XC5
4XC5
[84]
[85]
[86]
Monoterpenes
Sesquiterpenes
Diterpenes
[84]
[87]
Limonene
[85]
X2
[88]
Squalene

widespread in the lower terpenes, a second mode typifies the di- and triterpenes, namely, internal cyclization of one double bond onto another which has been polarized by an external electrophile ($H^\oplus$ or $HO^\oplus$). This is illustrated by the formation of manoöl [91] from geranylgeraniol, rewritten in a coiled form [90] to emphasize the ease of such ring formation (the polarizations of each olefin proceed in the natural, Markownikoff direction as well).

[70] ←——   ≡   ——→ [71]

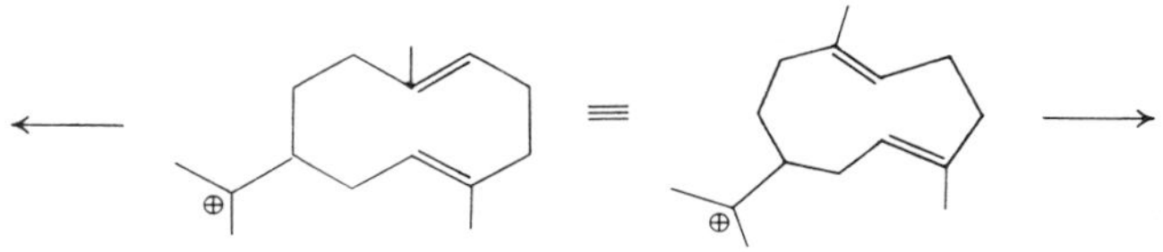

[89]

The triterpenes (C_{30}) and carotenoid pigments (C_{40}) arise, not from continuous polymerization of C_5-units, but from specialized dimerizations of the C_{15}- and C_{20}-pyrophosphates, [85] and [86]. The much more intensively studied reductive dimerization of [85] yields squalene [88], the fundamental triterpene and source of all the other triterpenes and steroids. Squalene is written in a coiled form in [88] to dramatize its potential for internal cyclizations of the manoöl kind, leading in this case to the steroids and triterpenes, as discussed in Chapter 4.

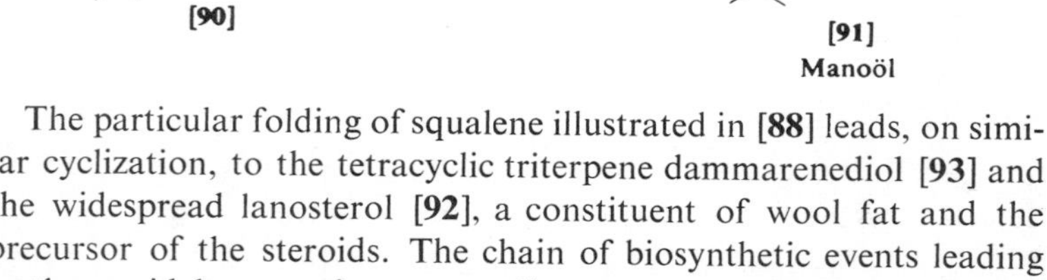

Cyclization
and allylic
rearrangement

[90]

[91]
Manoöl

The particular folding of squalene illustrated in [88] leads, on similar cyclization, to the tetracyclic triterpene dammarenediol [93] and the widespread lanosterol [92], a constituent of wool fat and the precursor of the steroids. The chain of biosynthetic events leading to these widely occurring tetracyclic compounds (Chart 8) has been extensively examined with radioactive tracers, particularly by Bloch and Cornforth, and includes stepwise oxidative removal of three methyls from lanosterol [92] to cholesterol [94], various changes in oxidation state at sites on the tetracyclic nucleus, and breakdown of the side chain from C_8 to C_5 (bile acids), C_4 (butenolide sapogenins), C_2 (cortical hormones); or none (sex hormones), as implied in the structures of Chart 8. The stereochemistry of the tetracyclic nucleus is always as shown in [94], which also includes the traditional numbering of positions. The lettered identification of the four rings is indicated in the structure of ergosterol [95], which also illustrates the C-24 methyl (sometimes ethyl) common to sterols found in plants, microorganisms, and marine invertebrates.

2–5   Biosynthesis of the Alkaloids

Usually found only in the higher plants, the alkaloids are certainly the largest (over 2500 compounds isolated to date) and most diverse of the families of natural compounds, and they contain the most complex molecular structures. Their diversity derives from the fact that the alkaloids are grouped as a family solely by the common presence of basic nitrogen in their structures rather than by the more detailed and restrictive biogenetic relations defining the previous two groups. Nevertheless, this fact is in itself of biogenetic significance, for the alkaloids arise from amino acids and, surprisingly, from only a few.

The biosynthetic pathways to the alkaloids were first worked out solely by inference from the existence of common structural features, a brilliant intellectual achievement begun over half a century ago by Pictet and Robinson. In recent times radioactive tracer work has generally confirmed their speculations and added much solid detail as well. We are now in a position to appreciate in broad outline the wonderful simplicity of Nature's design for the synthesis of these intricate molecules.

Most of the alkaloids contain aromatic rings and in many of these the structural unit [102] can be discerned. This was first recognized by Pictet, who further observed that the benzylisoquinoline alkaloids (cf. laudanosine [104]) contained two such units, which might be synthesized by a condensation such as [103]. Subsequently it was shown by Schöpf that such a condensation occurred with great facility *in vitro* at neutral pH and room temperature and so was a reasonable candidate for a biosynthetic pathway. Recently radioactive tracer work by Battersby and Barton has substantiated this in living plants.

As Robinson first observed, the reaction is in fact a generalized Mannich condensation and may be regarded as the condensation of an

CHART 8 ▪

Representative Examples of the Major Steroid Types

Methyl-steroids or
Tetracyclic Triterpenes:

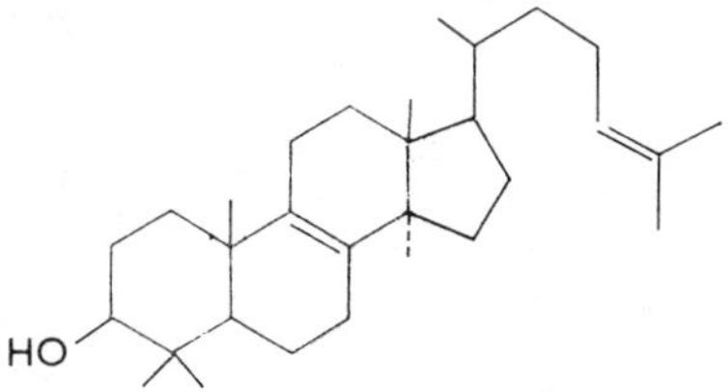

[92]
Lanosterol

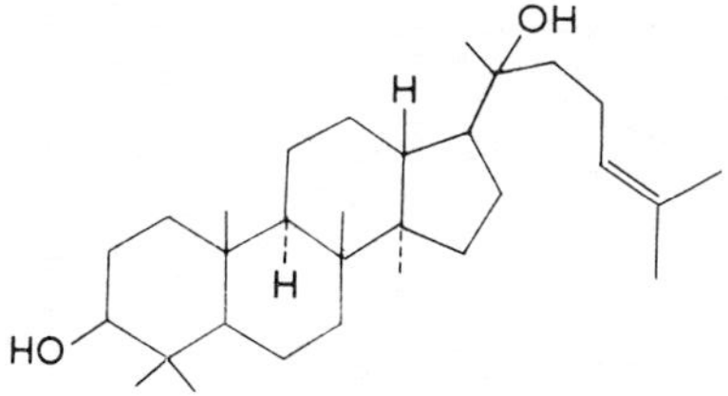

[93]
Dammarenediol

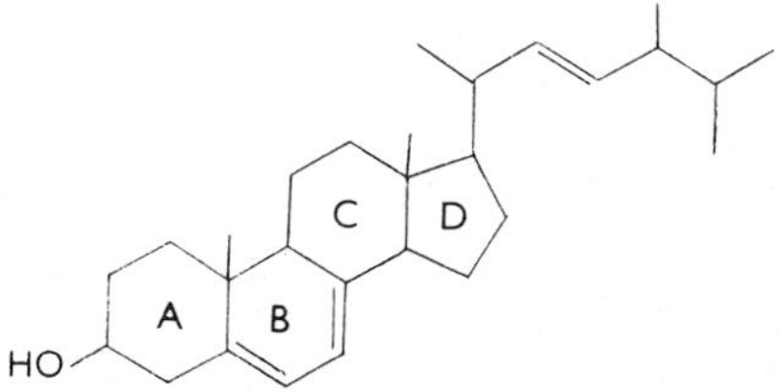

[95]
Ergosterol

Sterols:

[94]
Cholesterol

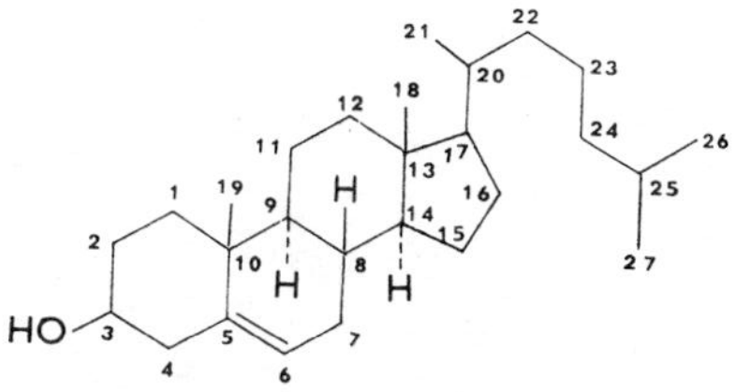

$$\left[\begin{array}{c} Bile\ Acids: \\ [4-83] \end{array}\right]$$

[97]

Digitoxigenin

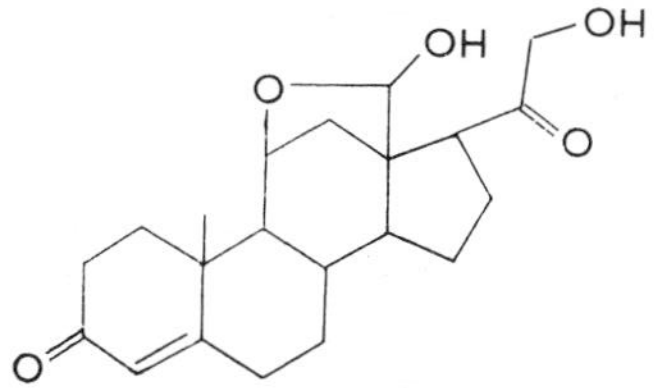

[99]

Aldosterone

Sapogenins:

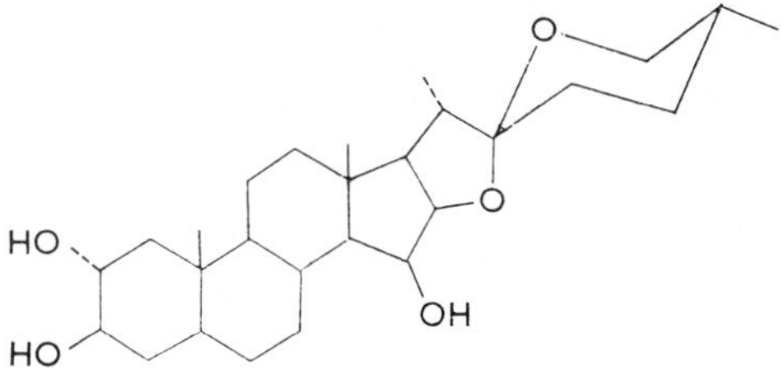

[96]

Digitogenin

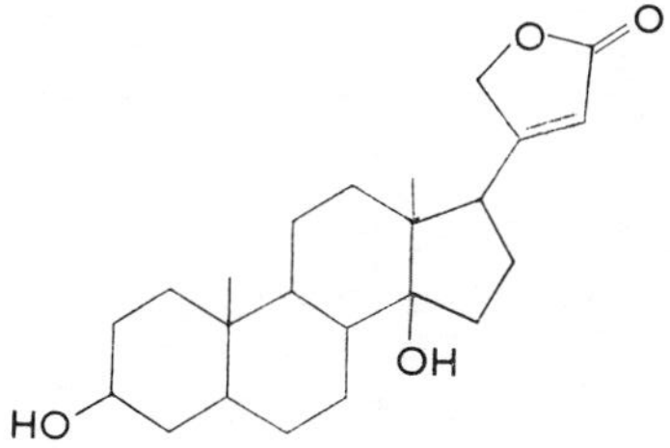

Adrenocortical Hormones:

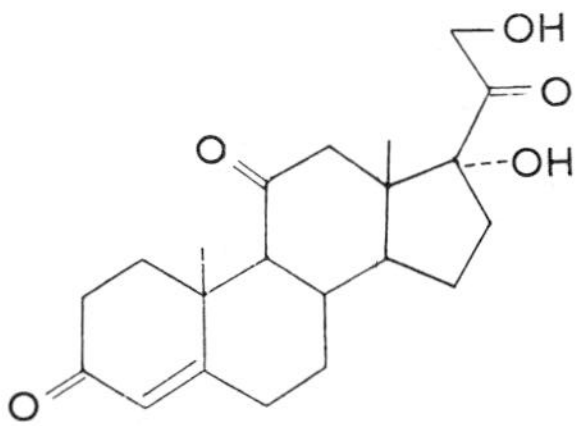

[98]

Cortisone

CHART 8 • (*Continued*)

Sex Hormones:

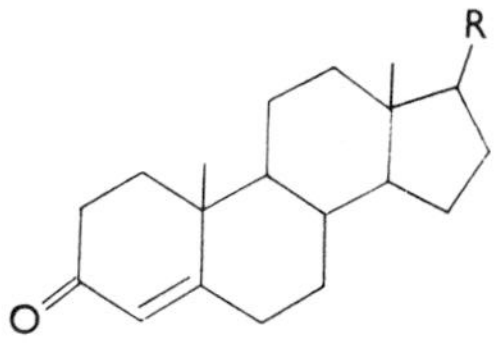

[100]

Testosterone, R = OH
Progesterone, R = COCH₃

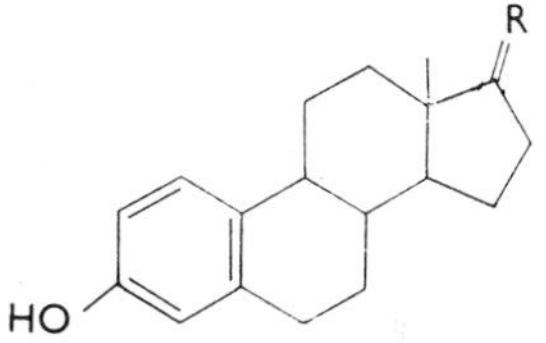

[101]

Estrone, R = O
Estradiol, R = H, OH

Alkaloids:
[4–98, 99, 100, 106]

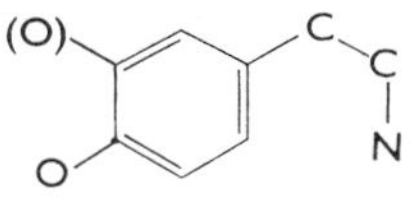

[102]

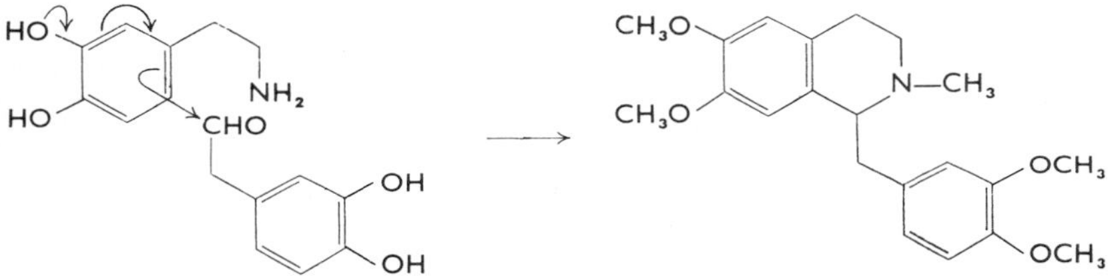

[103]

[104]

Laudanosine

aldehyde with an amine, yielding an imine (or immonium salt) which may then be attacked by a nucleophilic carbon such as that of an enol or phenol to form a carbon–carbon bond, as illustrated in [105] for an enol of a ketone, and implied for a phenol in the arrows in [103].

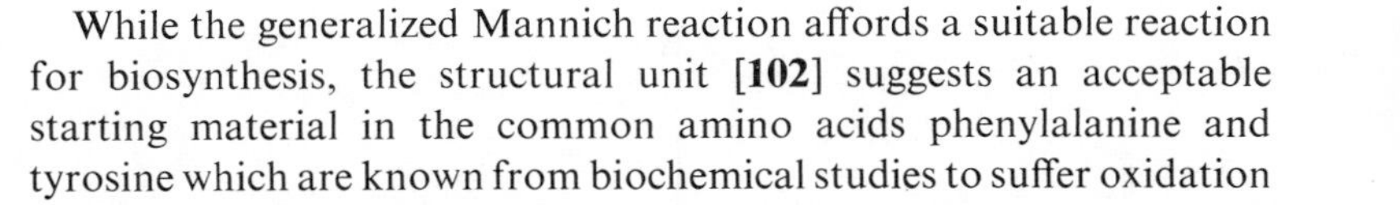

[105]

While the generalized Mannich reaction affords a suitable reaction for biosynthesis, the structural unit [102] suggests an acceptable starting material in the common amino acids phenylalanine and tyrosine which are known from biochemical studies to suffer oxidation to the *o*-dihydroxyphenyl derivative and decarboxylation to the amine, as well as oxidative deamination and decarboxylation to the aldehyde in common metabolic situations. The changes are summarized in Chart 9, which also lists the few amino acids involved in alkaloid biogenesis.

These, then, are the basic ingredients for the synthetic elaboration of alkaloid molecules in living tissue. The chief variants available in nature for affording diversity include the occasional involvement of small polyacetyl or ϕ-C$_3$ fragments as in the acetogenins and the

several reactions listed at the beginning of this chapter, particularly
N- and O-methylation, amine oxidation, and the oxidative coupling of
phenols at their ortho and para positions. These biogenetic schemes

CHART 9 ．

*Amino Acids and Their Transformations Involved in
Alkaloid Biosynthesis*

Alicyclic Group:

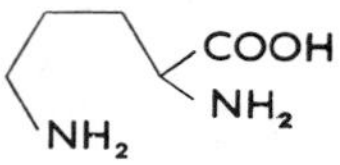

Ornithine

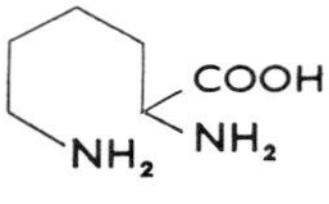

Lysine

Phenylalanine Group:

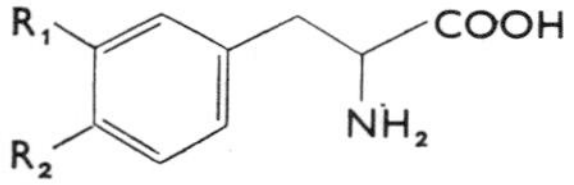

Phenylalanine: $R_1 = R_2 = H$
Tyrosine: $R_1 = H$; $R_2 = OH$
Dihydroxyphenylalanine: $R_1 = R_2 = OH$

Indole Group:

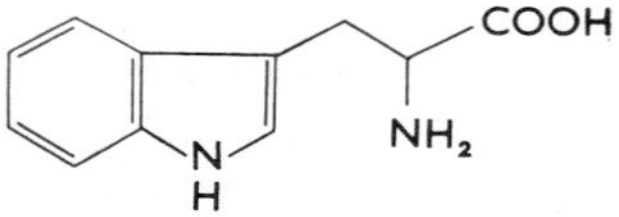

Tryptophan

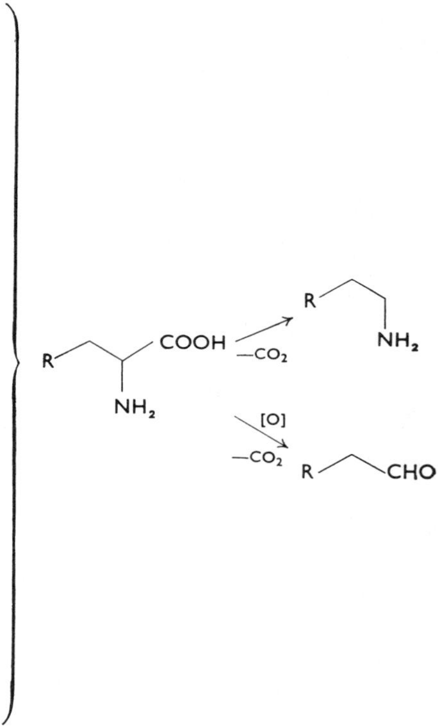

have been the inspiration for a number of elegantly simple laboratory
syntheses (cf. [103] → laudanosine [104]), several of which are included
in Chapter 5.

　　The amino acid sources of the alkaloids afford a convenient classi-
fication into three large groups, as indicated in Chart 9, and the

following introductory survey (Chart 10) of the structural families will be cast in this mold. The simplest alkaloids are those derived from the aliphatic amino acids (Chart 10a); their biogenesis was first outlined by Robinson (who regarded acetone-dicarboxylic acid, derived from citric acid, as a source we should now prefer to replace by a short acetogenin chain, viz., acetoacetic acid).

[106]

Hygrine [107] is typical of the pyrrolidines arising as implied in [106] from ornithine (with acetoacetic acid acting as the anion source for the Mannich reaction). Methylisopelletierine [110], from pomegranate root, is an example of the simple piperidines, the homologous structures deriving from lysine. Similarly, tropine [109] is an N-oxidized (to an imine) and further cyclized [106] hygrine with the ketone subsequently reduced. The carboxyl, lost in tropine, is to be seen, however, in cocaine [109], another tropane alkaloid retaining the full acetoacetate skeleton. The lysine analogue here is pseudopelletierine [112]. The aliphatic rings of the tobacco alkaloids nicotine [113] and anabasine [114] have been shown to arise from ornithine and lysine, respectively.

Two benzoylacetic (ϕ-C$_3$) units condensing with lysine afford the *Lobelia* alkaloids, such as lobelanine [111], which has been synthesized in the laboratory from benzoylacetic acid, methylamine, and glutaric dialdehyde. The many pyrrolizidine alkaloids in *Senecio* species (from two ornithine units) have the skeleton of platynecine [115], whereas the lupin alkaloids contain the homologous lupinine [117] from two molecules of lysine. The lupin alkaloids also include the more complex sparteine [5–6] and matrine [118], both of which may be constructed from three lysine-derived amino-aldehydes.

The phenylalanine group (Chart 10b) is of course characterized primarily by 1-benzylisoquinoline skeletons, although some simple isoquinolines are known, such as corypalline [119], which may arise simply by cyclization of an N-methyl oxidized to $> N^{\oplus}=CH_2$. Simple methylation affords the opium constituent laudanosine [104] which (with phenolic or methylenedioxy variants) may be regarded as the

CHART 10a ■

Representative Examples of the Alicyclic Alkaloids

Pyrrolidines:

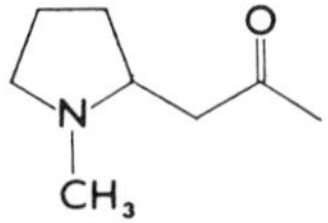

Tropane Alkaloids:

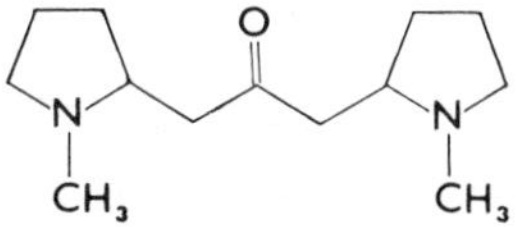

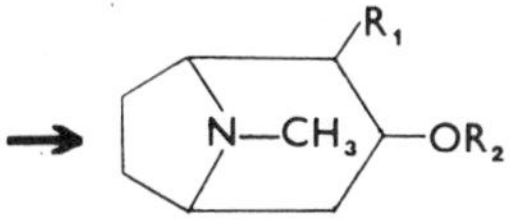

[109]

Tropine: $R_1 = R_2 = H$
Scopolamine: $R_1 = H$; $R_2 = \phi CHCO$— and epoxide at
Cocaine: $R_1 = $—$COOCH_3$; $R_2 = \phi CO$—

[107]
Hygrine

[108]
Cuskohygrine

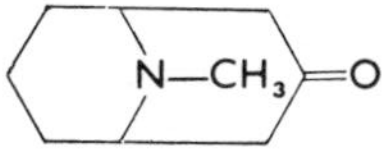

[112]
Pseudopelletierine

Piperidines:

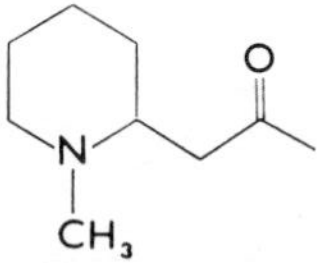

[110]
Methylisopelletierine

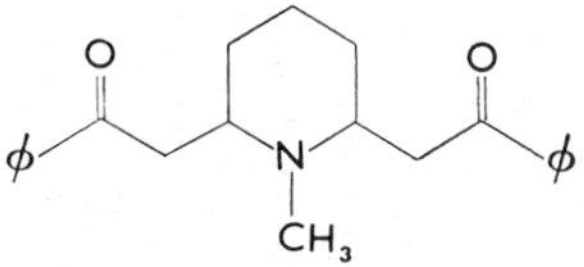

[111]
Lobelanine

Pyrrolizidine Alkaloids:

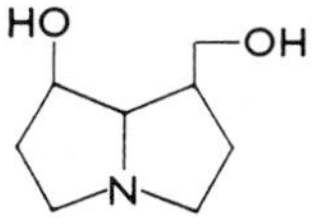

[116]
Junceine

[115]
Platynecine

Lupin Alkaloids:

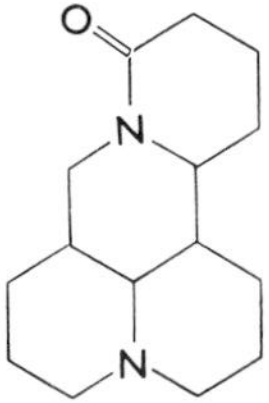

[118]
Matrine

[117]
Lupinine

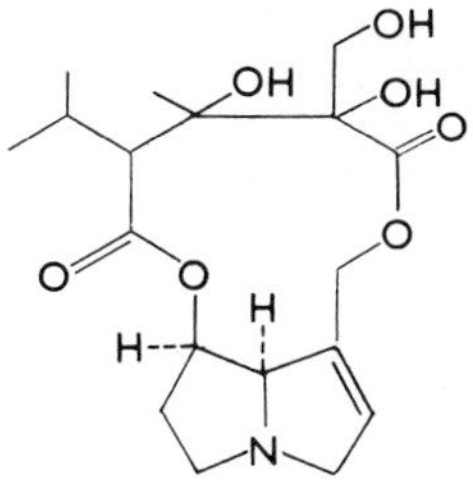

Tobacco Alkaloids:

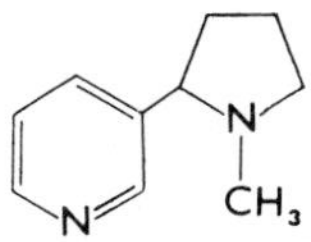

[113]
Nicotine

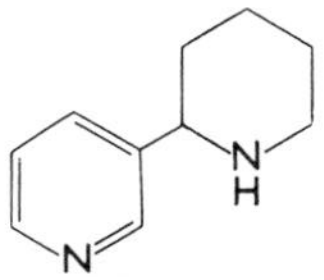

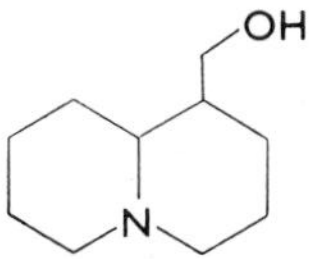

[114]
Anabasine

CHART 10b (PART I) ■

Representative Examples of the Phenylalanine Alkaloids

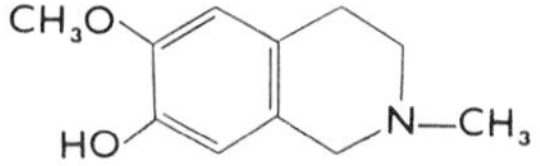

[119] Corypalline

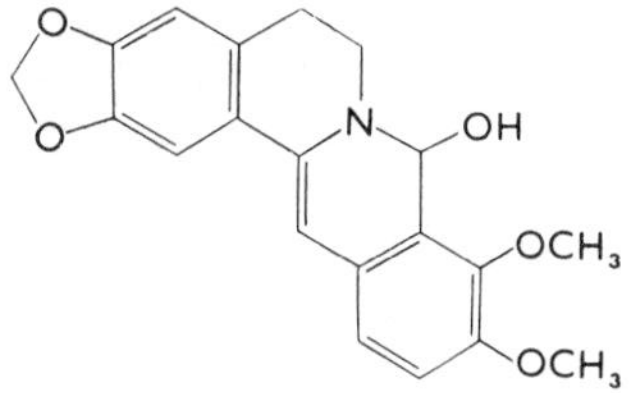

[120] Berberine

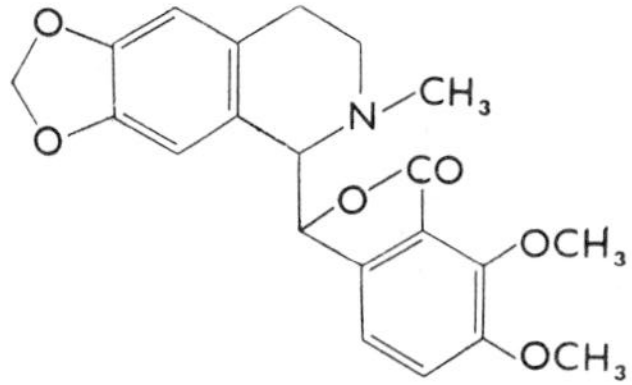

[121] Hydrastine

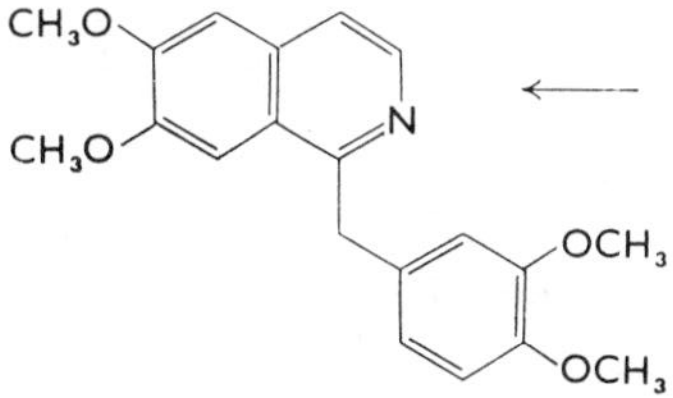

[122] Papaverine

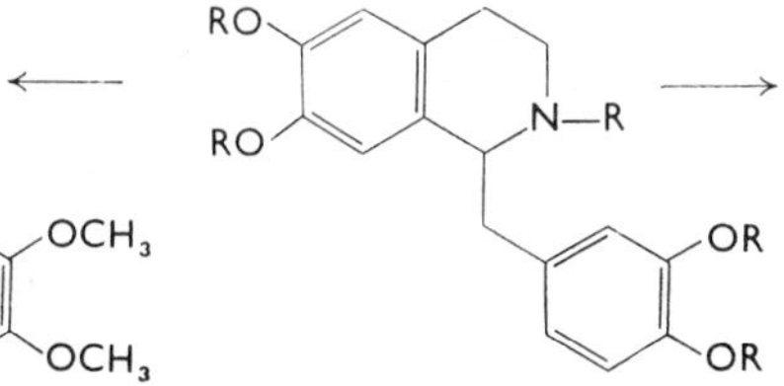

[123] Laudanosines (R = H, CH₃)

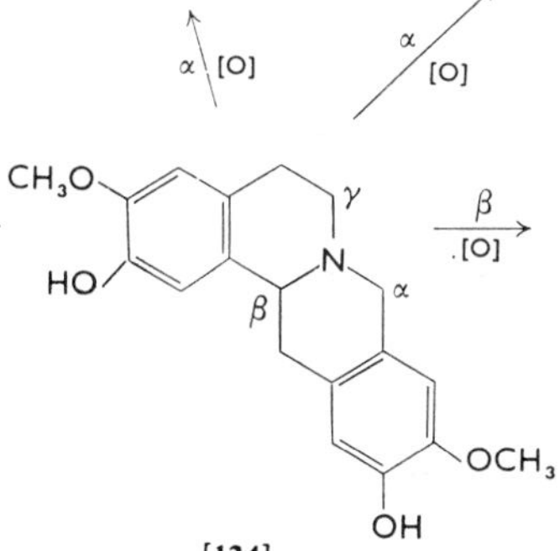

[124] Corexamine

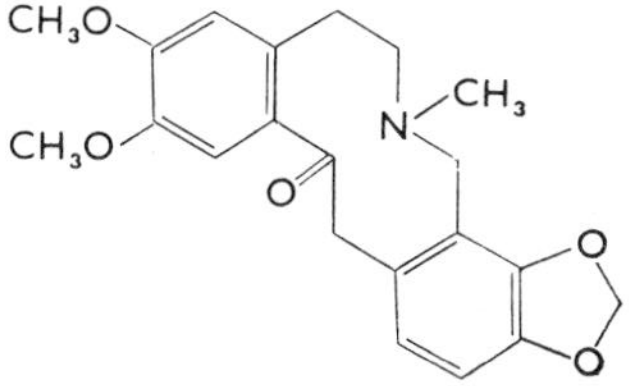

[125] Cryptopine

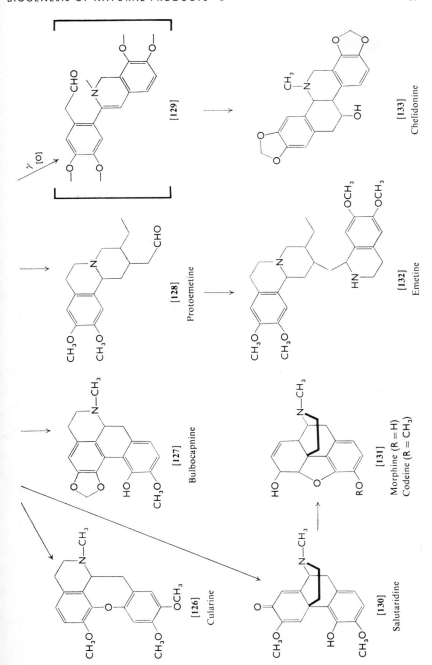

protean substance [123] for subsequent oxidations, after the several modes (outlined at the beginning of this chapter) which yield most of the main alkaloids of this group.

N-oxidation of a laudanosine can lead to aromatization of ring B as in papaverine [122] or to activation of the N-methyl for Mannich-type cyclization to the phenolic ring of the benzyl substituent, yielding the protoberberine group (cf. [124]). Further N-oxidation can proceed via each of the three carbons (α, β and γ in [124]) linked to the nitrogen. The α-oxidation can lead to aromatization as in berberine [120] or to ring cleavage as in the several phthalideisoquinoline alkaloids [121]. The β-cleavage with N-methylation creates the medium-ring amino ketones of the protopine class [125]. The γ-oxidation, cleaving (and another N-oxidation) to a hypothetical aldehyde [129], affords the benzophenanthridine family [133] by recyclization. Finally, cleavage between the phenolic oxygens of ring D of a protoberberine yields the skeleton found in the naturally occurring alkaloid [128], and a further Mannich condensation is easily envisioned leading to the emetine family [132].

Oxidative coupling of the two phenolic rings in the laudanosine-type precursor can proceed in three ways, via phenolic oxygen to the cyclic ethers (cf. [126]) or by carbon–carbon bond formation either ortho or para to the tyrosine phenolic position in ring A, producing either many aporphine alkaloids [127] or morphine alkaloids [131]. Tracer studies by Barton have confirmed this morphine biosynthesis in detail, the actual pathway involving the partially methylated salutaridine [130], itself an isolated natural product. Finally, oxidative coupling between the two molecules leads to the bisbenzylisoquinoline alkaloids, and the macrocyclic phenyl ethers such as tubocurarine [134], a major component of the South American curare used for poisoned arrows.

The erythrina alkaloids [136] also exhibit the quaternary center characteristic of morphine and may also be constructed biogenetically from an inner phenolic coupling of a precursor derived from condensing a phenylacetaldehyde and β-phenylethylamine. The cyclization of nitrogen apparently represents a second internal oxidation. Starting with a hypothetical imine from condensation of a shikimic-derived benzaldehyde and a β-phenylethylamine, we can readily construct the various families of *Amaryllidaceae* alkaloids by similar phenolic coupling. The precursor in methylated form is known in the alkaloid belladine [135], and ortho- or para-coupling produces the lycorine

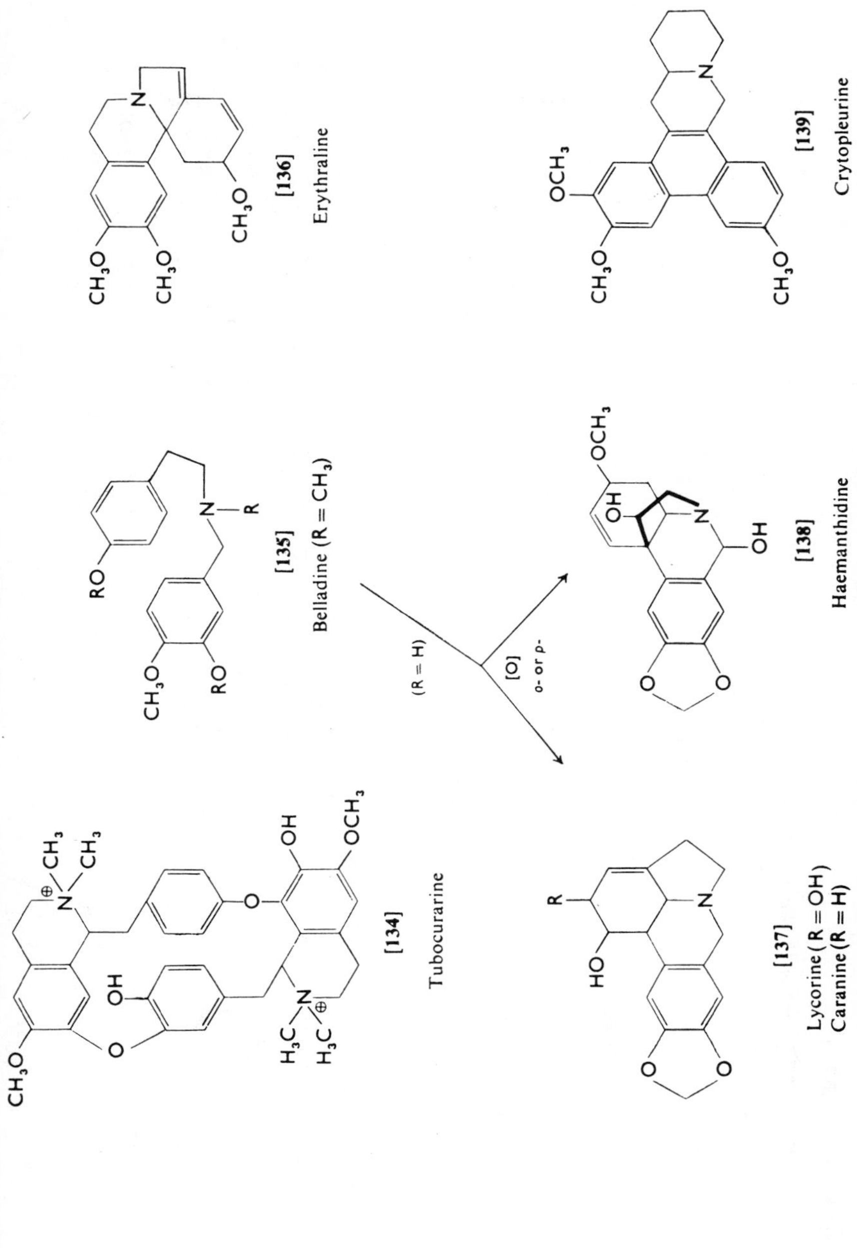

[136]
Erythraline

[139]
Crytopleurine

[135]
Belladine (R = CH₃)

(R = H)

[O]
o- or p-

[138]
Haemanthidine

[134]
Tubocurarine

[137]
Lycorine (R = OH)
Caranine (R = H)

CHART 10c (PART I) •

Representative Examples of the Indole Alkaloids

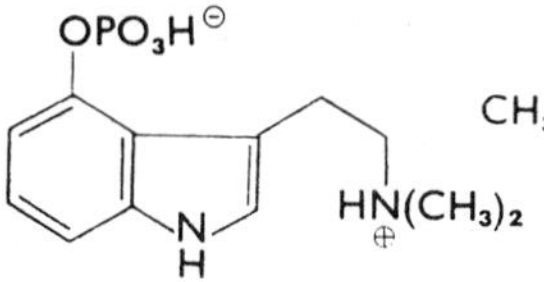

[140]
Psilocybin

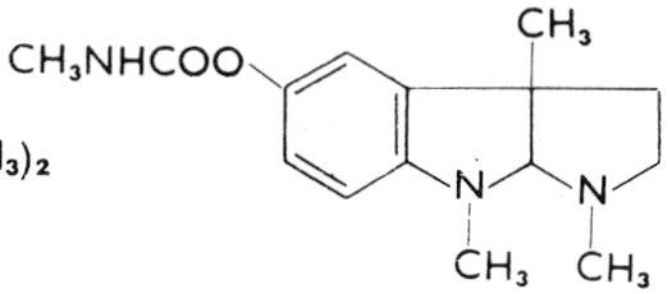

[141]
Physostigmine

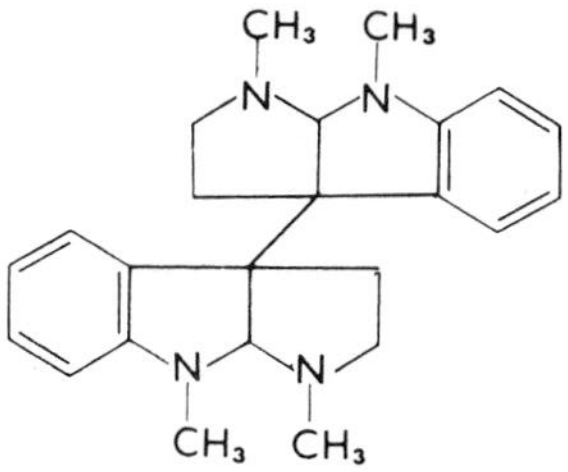

[142]
Folicanthine

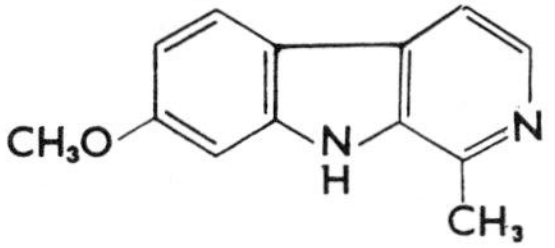

[143]
Harmine

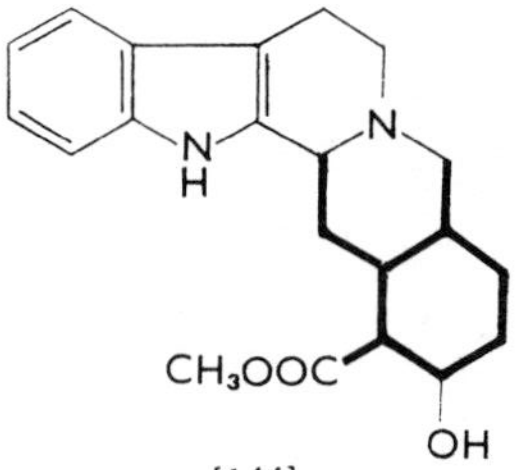

[144]
Yohimbine

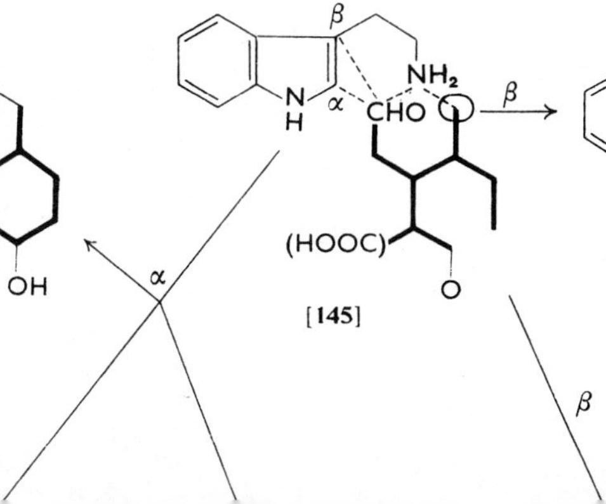

[145]

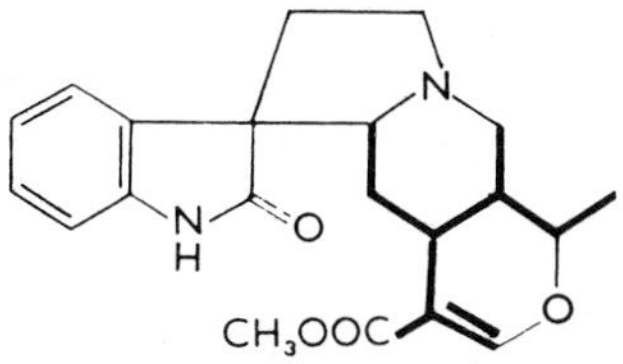

[146]
Mitraphylline

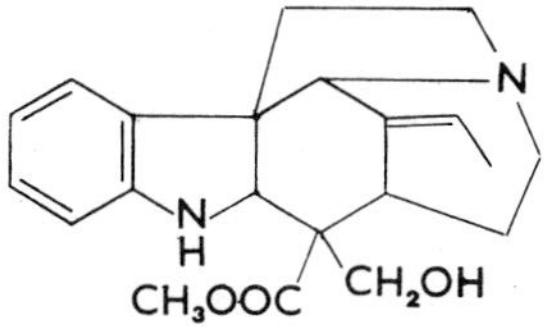

[147]
Stemmadenine

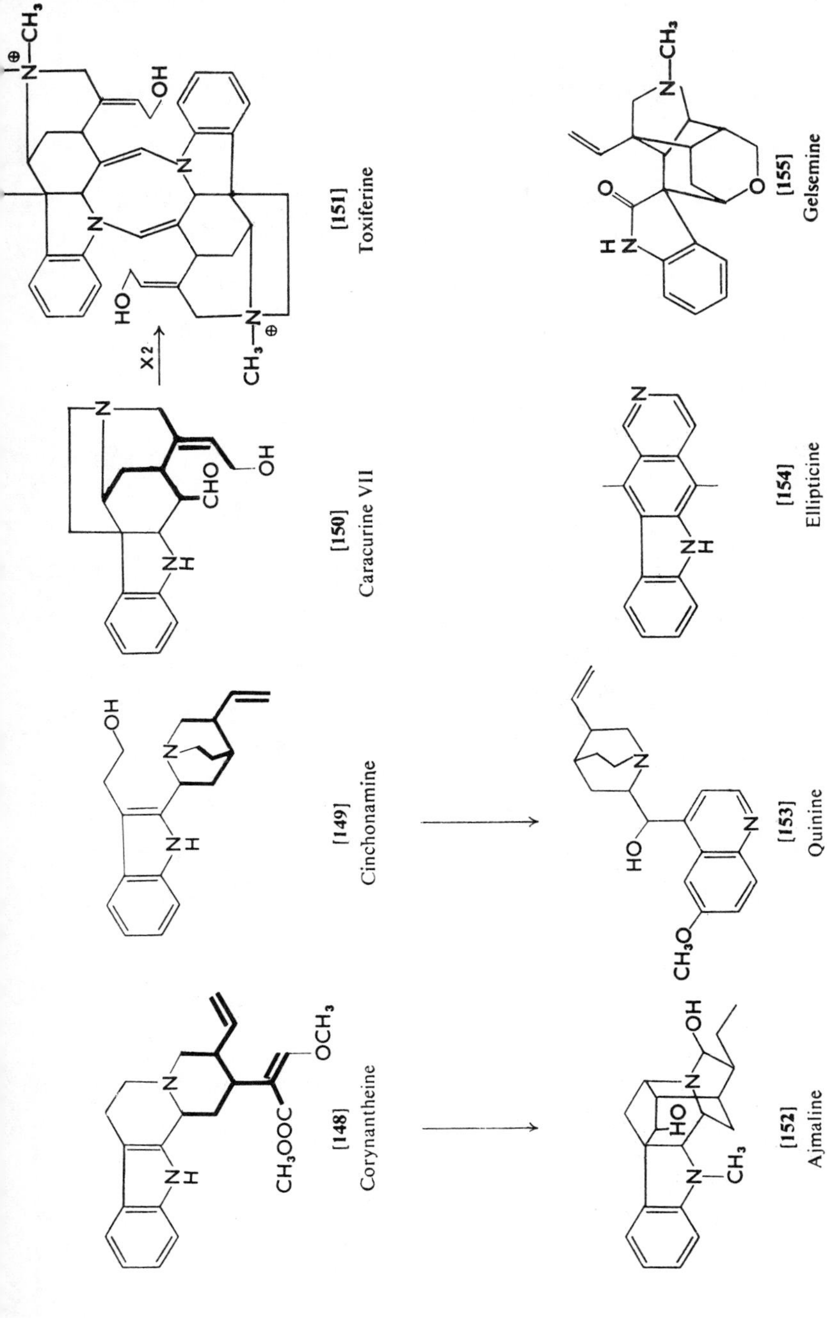

[151]
Toxiferine

$x^2 \longrightarrow$

[150]
Caracurine VII

[149]
Cinchonamine

[148]
Corynantheine

[155]
Gelsemine

[154]
Ellipticine

[153]
Quinine

[152]
Ajmaline

CHART 10c (PART II) ■

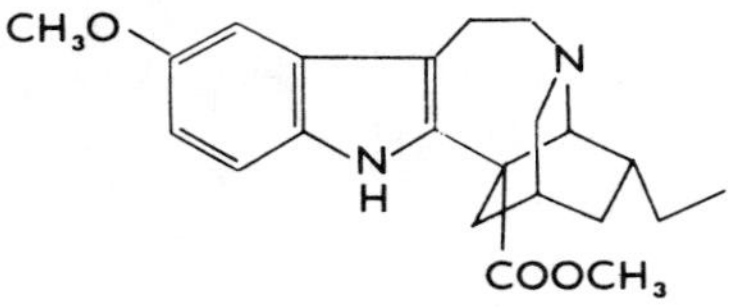

[156]

Voacangine

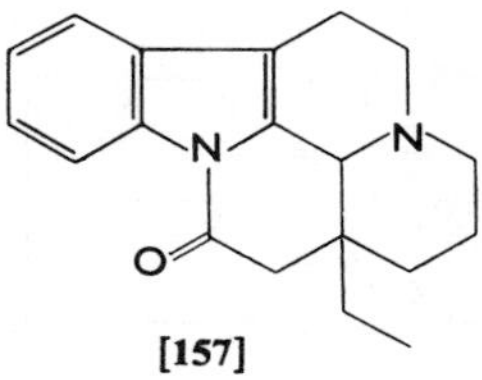

[157]

Eburnamonine

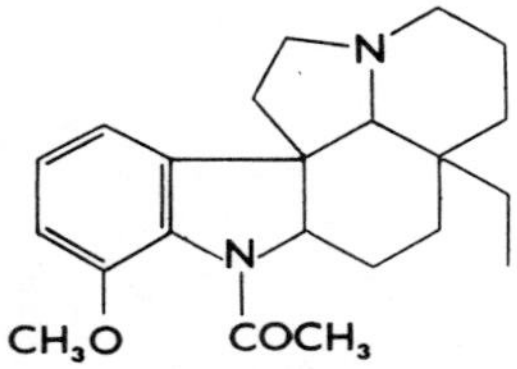

[158]

Aspidospermine

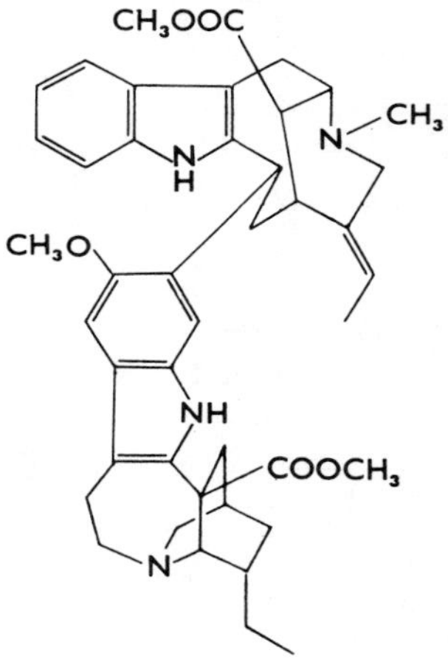

[159]

Voacamine

CHART 10c (PART II) ▪ *(Continued)*

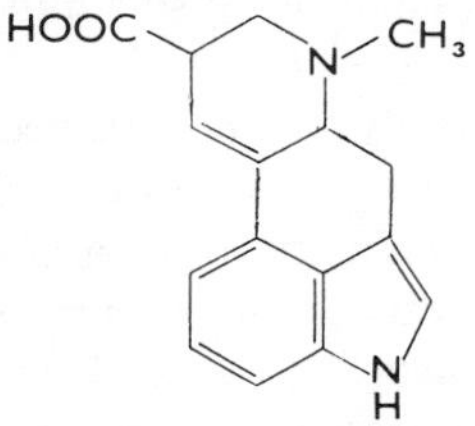

[160]
Lysergic Acid

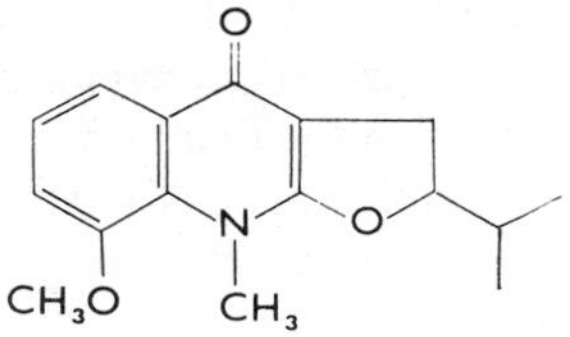

[161]
Lunacrine

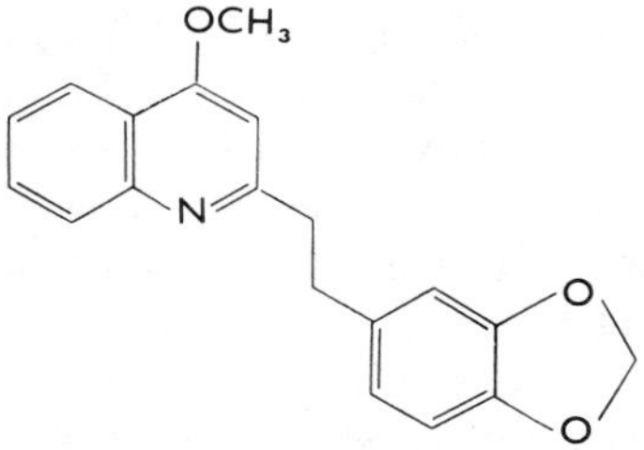

[162]
Cusparine

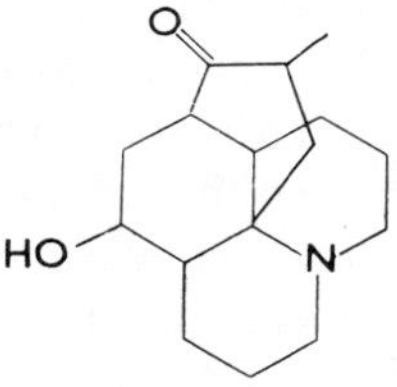

[163]
Annofoline

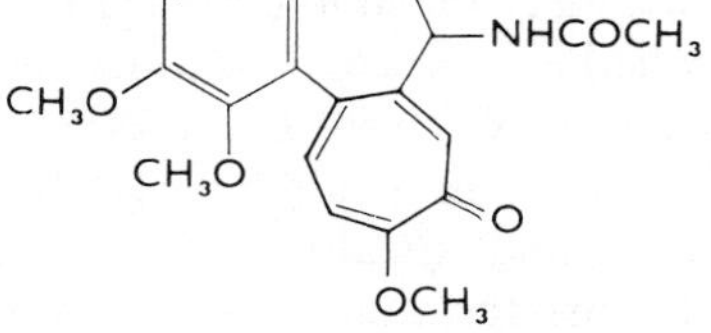

[164]
Colchicine

[137] and crinine ([5–45]) families, corresponding to aporphine and morphine formation, respectively.

The interesting skeleton of cryptopleurine [139] apparently represents a biogenesis from two phenylalanines and one lysine unit, a rare link between the two families.

The indole group (Chart 10c) is structurally the most varied of the three and also retains at present some biosynthetic mysteries still under active tracer study. Simple tryptamines such as psilocybin [140], the hallucinogenic principle of the Mexican ceremonial mushroom, are found in nature. The indole ring is mechanistically available for alkylation or oxidation both at the α- and β-positions; β-methylation (and spontaneous cyclization of the amine) of a tryptamine can thus lead to physostigmine [141], an African ordeal poison; β-oxidative coupling yields the *Calycanthaceae* group (cf. folicanthine, [142]). The Mannich condensation (with an equivalent of CH_3CHO) proceeds at the α-position to fashion the harmine group [143].

By far the largest group of indole alkaloids is derived from tryptamine by Mannich condensation either at the indolic α- or β-positions with a nine-or-ten carbon fragment of unknown origin [145]. Originally believed to be a phenolic phenylacetaldehyde (capable of a fission between the 3,4-dioxy sites similar to emetine [128]) as in the phenylalanine-derived alkaloids, this fragment has since been shown in a number of cases not to arise from labeled tyrosine and like precursors. Many of these indole alkaloids contain the skeletal fragment biogenetically symbolized in [145]. They are shown in heavy lines in some of the examples but in some molecules, particularly those from *Tabernanthe* [156] and *Hunteria* [157] species, and the aspidospermine [158] family, they are rearranged. The circled carbon in [145], as in berberine, may have its origin in an oxidized N-methyl.

The umbrageous fragment, regarded as an aldehyde [145] for Mannich condensation to the indole, leads to two groups of alkaloids depending on whether its condensation proceeds at the α- or β-position of the indole. Cyclization involving the α-position usually leads to true indoles, as in yohimbine [144], corynantheine [148], cinchonamine [149], and ajmaline [152]; whereas β-condensation must block indolic unsaturation with a quaternary center as in mitraphylline [146], caracurine VII [150], and strychnine [5–93].

A number of other alkaloids are biogenetically derived from these by a few further transformations, as in the conversion of cinchonamine (or a close relative) to the quinine [153] family skeleton via opening of the indole ring and reclosure of the aniline nitrogen on the side chain. Indeed, Leete finds radioactive tryptophan a precursor of this group. Interesting if somewhat more involved schemes may be constructed to afford stemmadenine [147], ellipticine [154], and gelsemine [155] from a starting point of the caracurine [150] type. The hallucinogens from the rye fungus (ergot), thought to have been the cause of St. Vitus's Dance in medieval times, are peptide amides of the interesting lysergic acid [160], shown by tracer studies to be a biosynthetic product of tryptamine and mevalonic acid! The most complex bases are the dimeric indole alkaloids such as the many constituents of calabash curare, derived from caracurine by a simple dehydrative dimerization and exemplified in the common toxiferine [151]; and voacamine [159] and the very similar vinblastine, an anti-cancer agent isolated from the humble periwinkle.

Finally, a number of quinoline alkaloids such as lunacrine [161] and cusparine [162] (the major component of angostura bitters), traditionally regarded as originating from anthranilic acid but very possibly derived directly from tryptophan, may be for organizational purposes included with the indole alkaloids. Apart from these three major groups of alkaloids, there is also a well-defined group of terpenoid alkaloids (Chapter 4, pp. 118, 135) and a smaller group of acetogenin alkaloids [163], compounds in which the skeleton may be recognized as belonging biogenetically to these other main families. The autumn crocus elaborates the remarkable alkaloid colchicine [164], a mitotic poison also successfully used in the treatment of gout.

3

ACETOGENINS AND
PHENYLPROPANES ▪

THE BRILLIANT COLORS that surround and delight us in the natural world
generally consist of three kinds of pigments: the tetra-pyrrol group,
chiefly heme and chlorophyll; the terpenoid carotenes; and the acetogenins. The last group is by far the most numerous and varied, both in
color and chemistry. There can be little doubt that the study of these
compounds loomed large in the early years of organic chemistry
because of their esthetic attraction and the relative ease of following the
passage of colored material through extraction and purification
procedures. Acetogenins are so widespread in natural pigmentation
that they account for most flower colors and autumn foliage as well as
the deep hues of aphids and sea urchins; of rhubarb and multicolored
lichens; of vivid tropical woods as well as molds and fungi.

Well over three-quarters of the thousand or so known acetogenins
are polyphenols, and, as most of these also have conjugated carbonyl
groups as well, it is no surprise that their absorption spectra have

maxima in the visible region. The names of many of the oxygen-heterocycles so characteristic of the acetogenins reflect their colored nature: xanthones, flavones, chromones, anthocyanins. The biogenesis of these compounds from the polyacetyl chain is generally clearly recognizable in their structures and their behavior and allows us to make some important generalizations about their chemistry.

3–1   Survey of Chemical Behavior; Flavasperone

The main observation is that the reactions of carbonyls, and most particularly β-diketones, are the prime characteristic of most of these molecules. Although tautomerism to enols (and phenols) and subsequent etherification of these may be exhibited, still these groups often react as ketones and their disguises should always be recognized. The most elaborate structural disguise, which is very widespread, is the appearance of oxygen-heterocycles[1] from cyclization of the ketone oxygens of the polyacetyl chain either to the chain itself or to the common isoprenoid substituents. Many examples of this are apparent here and in Chapter 2. It is important to remember that these complex heterocycles are often vulnerable to reagents which normally attack ketones.

The poly-β-ketone nature of most acetogenin molecules accounts for the fact that the most common reagent for their structural degradation has been hydroxide ion, which readily cleaves β-diketones. As a result the oxidative (selenium dehydrogenation) and reductive (zinc distillation) approaches common to the elucidation of the skeletons of terpenes and alkaloids are much less in evidence here.

An overall survey of the acetogenins reveals other differences from terpenes and alkaloids, most of which stem from the usually almost fully aromatic nature of acetogenin molecules. Thus, though appearing complex, they are often simpler to elucidate since they rarely exhibit the complications of stereochemistry. The chief chemical subtlety often lies instead in their capacity for tautomerism. Being aromatic, they also present more opportunities to bring to bear the power of physical methods. The most obvious and earliest were UV and visible spectra since each of these aromatic systems has a distinct and characteristic chromophore while their common enolic character allows more salient information to be drawn from the spectral shift in base as the

---

[1] The main oxygen-heterocyclic systems commonly found in acetogenins are assembled in Chart 1 with their characteristic numbering.

enol or phenol is converted to its anion. Similarly, the wide range of chemical shifts for olefinic and aromatic protons allows well-spread and clearly defined NMR spectra, in sharp contrast to the common overlapping blur of signals in saturated molecules such as steroids. Furthermore, whereas alkaloids are the natural bases, most aceto-genins are acids; in both groups acidity measurements frequently offer

CHART I ∎

The Main Oxygen-heterocycles in Natural Products

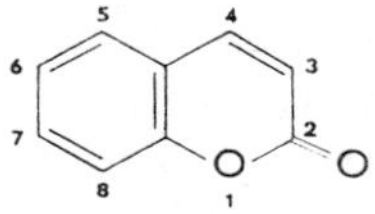

Coumarin	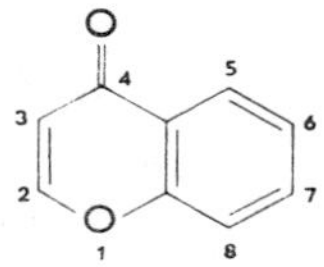 Chromone	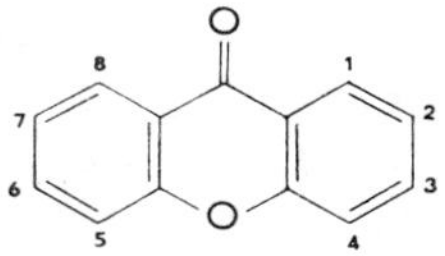 Xanthone
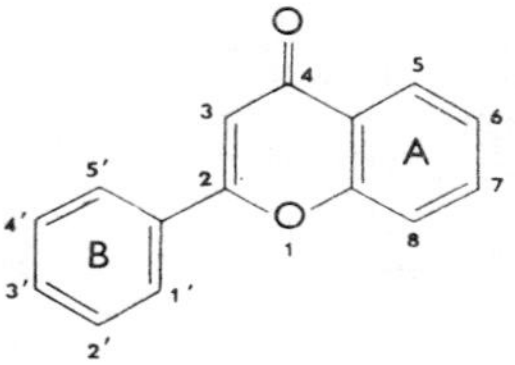 Flavone	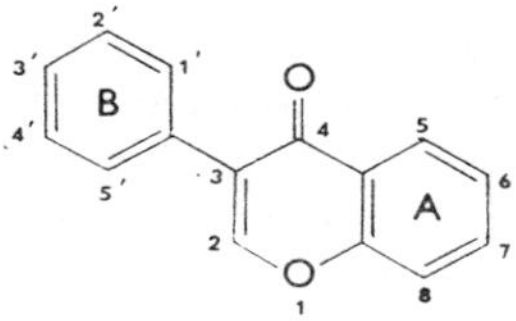 Isoflavone	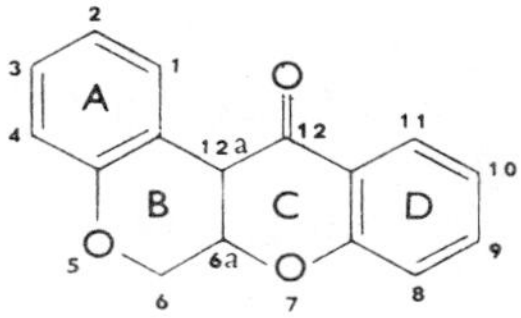 Rotenoid
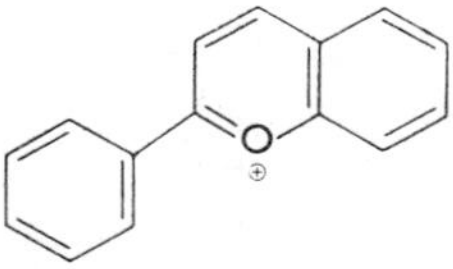 Anthocyanidin	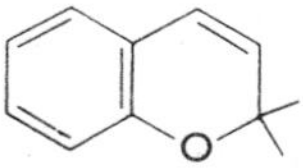 Dimethylchromene	Depsidone

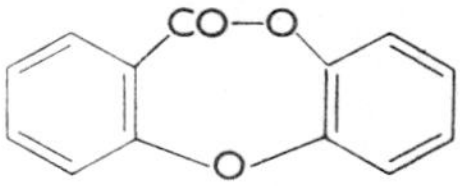

significant structural information. On the other hand, the weak basicity of certain oxygen heterocycles among the acetogenins may also be useful, most commonly in the formation of pyrilium salts from γ-pyrones (cf. chromones and flavones) in concentrated acids.

A few more degradative tools have had widespread applicability in this area. The enolic (or phenolic) nature of the compounds has traditionally been appraised with the $FeCl_3$ test. Other color tests were much more common in the years before instrumental methods were available, but were less reliable and are rarely used in contemporary work.

Polyphenols are readily oxidized, especially in base. This meant that oxidative degradations on the phenols were usually tried, although it was also common to isolate only very small fragments in this way (formic, oxalic, and acetic acids, for example). On the other hand, the fully methylated phenol ($(CH_3)_2SO_4/^{\ominus}OH$) is usually resistant to oxidative incursion into the benzene ring, and so frequently offers more significant products with that ring intact.

The methylated phenols are often seen in structure studies, and in those cases in which some phenolic groups are already methylated in the natural compound, the free phenols which remain are marked by using ethyl sulfate instead of methyl in the degradations. Acetylation or methylation serves to "count" the free phenolic groups, using the quantitative acetoxyl or Zeisel methoxyl determinations, and the presence of nonphenolic hydroxyl groups may be deduced by their reactivity to Ac_2O but not to $(CH_3)_2SO_4$. Finally, many acetogenins are quinones as evidenced by their easy reversible reduction-oxidation and by reductive acetylation.

The ketonic tautomer of a phenol is virtually totally suppressed by the resonance energy of the benzene ring, which easily overcomes the ordinary thermodynamic preference for the keto over the enol form. Nevertheless, that small amounts of ketonic tautomer can exist in a base-catalyzed situation is clear from such reactions as the decarboxylation of salicylic acid or the retroaldol cleavage of o-hydroxybenzyl alcohols, liberating aldehydes. The acetogenins are commonly resorcinols or phloroglucinols, the meta-orientation of their hydroxyls echoing the polyacetyl chain of their origin. In these phenols the thermodynamic barrier to formation of catalytic amounts of ketonic tautomer is much lower owing to the near-cancellation of the aromatic resonance energy by the unfavored enolic forms of two or three ketone groups on the ring. Hence in these compounds reactions involving the intermediacy of the keto tautomer are much easier.

An illustration is provided by the alkaline degradation used in the structure elucidation of flavasperone [1]. The cleavage of enol (and phenol) ethers by base involves Michael addition of $^{\ominus}OH$ followed by elimination of $^{\ominus}OR$ and is implicit in the first two stages [1]→[3]. Subsequently, the β-diketone (or vinylogous β-diketone) cleavages (dotted lines in [3]) serve to provide orcinol [4]. Not only is a variety of the base-catalyzed transformations common to many acetogenins in evidence here, but the example also serves to show how misleading these subtle reactions can be for structural inference. Assuming the intact C-methyl of orcinol to be present in the natural product, the original investigators first assigned flavasperone the structure [5].

The polyacetate biosynthesis of these phenolics is now so well established, however, that this structure would today have been regarded with grave suspicion since it cannot be derived from the linear polyketo chain and the rules of Chapter 2. (As if the basic degradation were not troublesome enough, HI demethylation yields nor-rubro-fusarin ([6], R = H)! Rubrofusarin ([6], R = CH_3) is another natural fungal pigment.)

3–2   Flavonoids

The largest group of acetogenins is the flavonoids, a series of plant and flower pigments with the skeleton of flavone (Chart 1). This group is varied most commonly by the oxidation state of the heterocyclic ring and the pattern of oxygen substituents on the nucleus: usually hydroxyl, methoxyl, or glycosidyl. These ubiquitous and delightful dyes of the plant world have provided a rich field of scientific investigation for decades, commencing with the first systematic studies of Kostanecki at the end of the nineteenth century and flowering in the first decades of the twentieth century, notably in the laboratories of Perkin and Robinson, where many natural products chemists, later more famous for other work, first contributed to the chemistry of flavonoid pigments. These compounds provide an appropriate introduction to the natural polyphenols.

Their varicolored metal chelates, arising from hydroxyls adjacent to the carbonyl, were once valuable both for commercial dyeing and for color-test recognition of various flavonoid types. Now we may most usefully recognize them by their UV spectra, the flavones characteristically exhibiting two strong absorption bands at 250–270 and 300–380 mμ depending on hydroxyl substitution. Furthermore, these bands are shifted in particular ways on the addition of strong or weak bases, chelating agents or Lewis acids; the spectral shifts can go far not only in identifying an unknown as a flavone but also in establishing the nature and position of many of its oxygen substituents!

The fundamental degradation procedure developed by Kostanecki consists of alkaline hydrolysis—usually on the methylated phenol—which opens the chromone ring to a β-diketone. This in turn is cleaved in each of the two possible ways, affording a mixture of two acids and two ketones which are readily identified (originally by synthesis). The procedure is shown for rhamnetin [7], combined with the ethylation technique for distinguishing which of the original oxygens were phenolic. The two phenolic products, from ring A, may proceed further under stronger alkaline conditions to lose CO_2, or a substituted acetic acid, respectively, by alkaline cleavage at the C-4 carbonyl. Since the phenolic products arise only from ring A owing to release of the chromone ether as a phenolic-OH, this degradation is sufficient to write the formula of the flavone.

This procedure is one which puts a premium on synthesis of the degradation products to confirm their identity and of the flavone itself. The synthetic methods for flavones developed over the years are all founded in β-diketone chemistry. Originally, the Claisen condensation of an appropriately methoxylated acetophenone with a benzoic ester [8] led to a β-diketone [9], demethylation of which with HI led to spontaneous cyclization to the flavone [10]. To avoid demethylation,

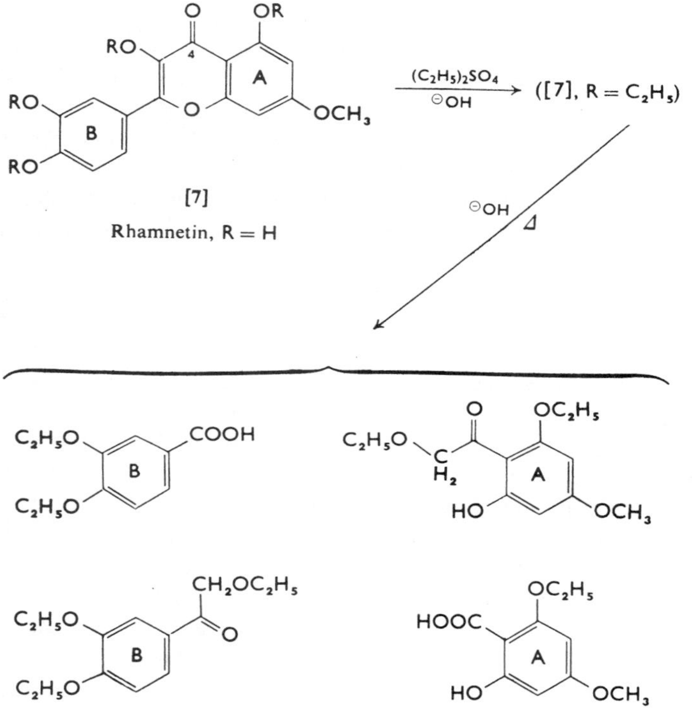

[7]

Rhamnetin, R = H

the acid component may be esterified to the *o*-hydroxy group of an acetophenone ([8], R = H), and the Claisen condensation carried out intramolecularly. Performic acid oxidation of the intermediate β-diketone [9] leads to the common 3-hydroxy-flavones ([10], R = OH).

The steps may all be combined by heating an *o*-hydroxy-acetophenone with both the sodium salt and the anhydride of the acid which is to be the source of ring B; the flavone is isolated thereby in one step. Finally the condensation of the *o*-hydroxy-acetophenone with an appropriate (for ring B) benzaldehyde yields a chalcone (φCH=

CHCOϕ) easily and this, in turn, cyclizes readily to a flavanone (2,3-dihydroflavone). Flavanones are dehydrogenated to flavones by SeO_2 or oxidized to 3-hydroxy-flavones by the expedient of nitrosation (RONO/OR$^\ominus$) to the 3-oxime and acid hydrolysis.

We owe the vivid red-to-blue spectrum of most flower colors to their anthocyanin content. The natural pigments are glycosides (usually at the 3-position) of the flavylium cations (anthocyanidins), commonly hydroxylated as in cyanidin [11], delphinidin [11], and petunidin [11]. Their names often refer to the flowers from which they were first obtained. The glycosides are soluble in water or alcohols and often isolated by soaking the fresh-picked flowers overnight in weak methanolic HCl, filtering the solution and then precipitating the blue lead salts by addition of aqueous lead acetate. Treatment of the lead salt with stronger methanolic HCl and filtering allows subsequent precipitation of the red anthocyanin chloride on the addition of ether. Slowly raising the pH of anthocyanin solutions causes a spectacular profusion of color changes owing to abstraction of protons from ring A hydroxyls.

The isoflavonoids are so similar in properties to the flavonoids that cursory assignments of structure have not infrequently had to be revised when synthesis proved them wrong. As anticipated on mechanistic grounds, of course, alkaline hydrolysis yields formic acid and a desoxybenzoin. The reverse route is the accepted synthesis but is effected most cleanly by reaction of the desoxybenzoin [12] with ethoxalyl chloride in pyridine to yield an α-keto-ester [13], in which the phenolic groups are also automatically protected; the intermediate readily cyclizes to a 2-carboxy-isoflavone ester [14], which may be saponified under very mild conditions and decarboxylated. The synthesis of genistein [15] is shown as the example.

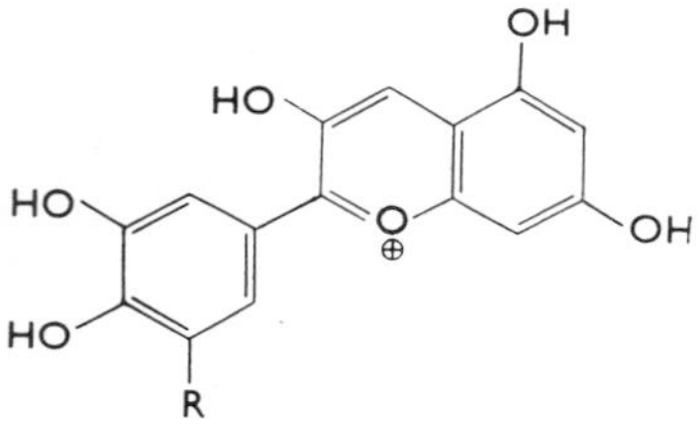

[11]

Cyanidin, R = H
Delphinidin, R = OH
Petunidin, R = OCH₃

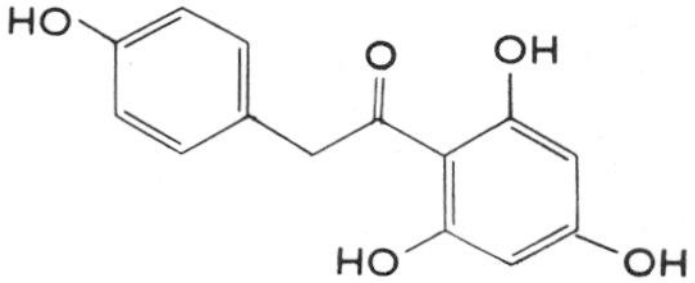

[12]

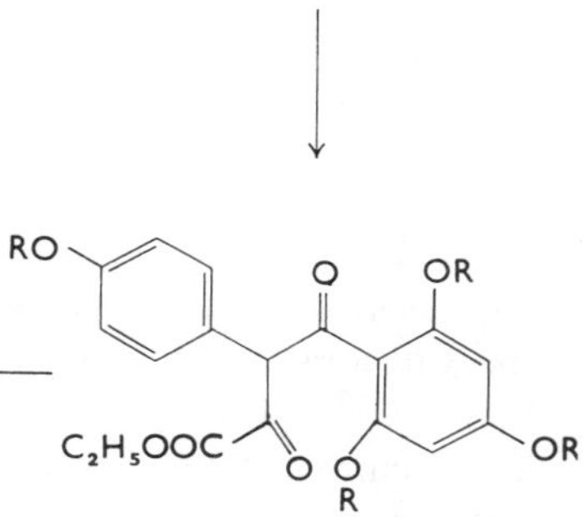

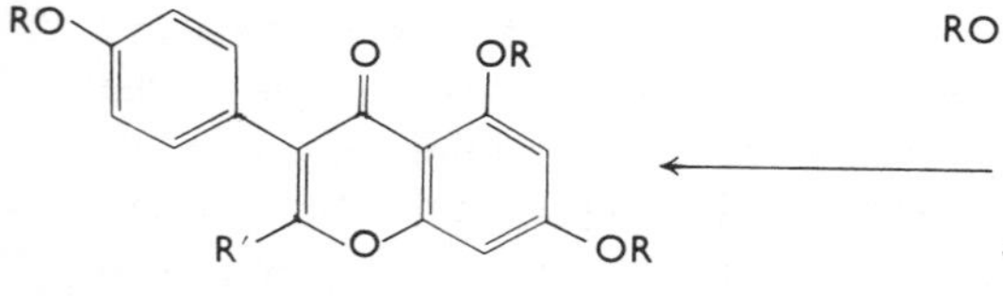

[14]: R = COCOOC₂H₅ ; R′ = COOC₂H₅
[15]: Genistein, R = R′ = H

[13]: R = COCOOC₂H₅

We owe our knowledge of the biosynthesis of flavones and iso-flavones largely to Grisebach, who fed radioactive precursors to red cabbage seedlings and degraded the cyanidin [11] by the normal alkaline hydrolysis. The C^{14}-acetate caused labeling of ring A (isolated as phloroglucinol and degraded by Birch's procedure, p. 81), and labeled phenylalanine yielded radioactive 3,4-dihydroxybenzoic acid and inactive phloroglucinol. In three parallel experiments using phenylalanine labeled with C^{14} in each of the three side-chain positions, the isoflavone formononetin [16] was isolated from red clover, purified to constant radioactivity, and degraded as shown in Chart 2 to give the percentages of original phenylalanine activity indicated, and thus to demonstrate the origin of isoflavones by phenyl migration in flavone biosynthesis.

CHART 2 ▪

Isoflavone Biosynthesis; Tracer Studies

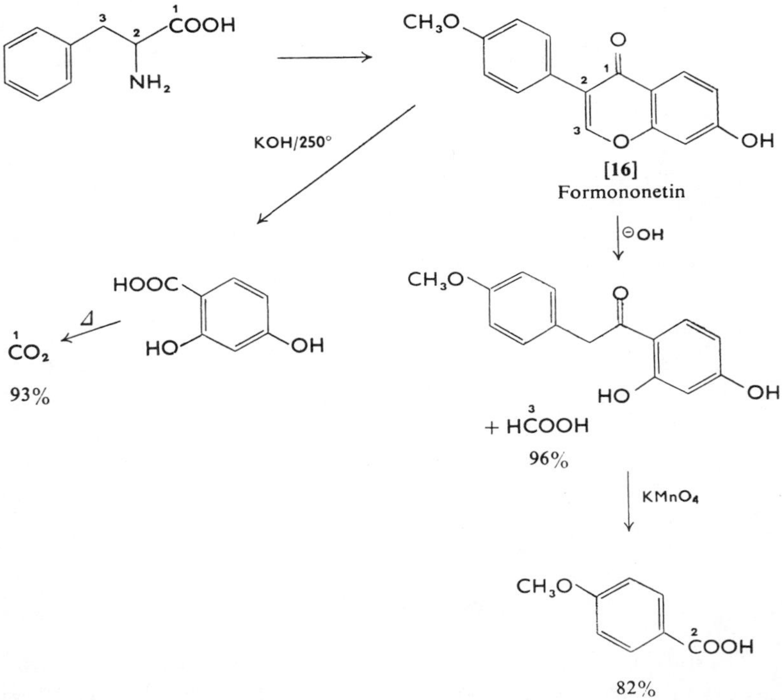

[16]
Formononetin

93%

+ H$\overset{3}{\text{C}}$OOH
96%

82%

3-3 Natural Pyrones

The fundamental heterocycle of the flavones is a γ-pyrone, and simple α- and γ-pyrones are both found in nature. Their properties are very similar, especially in the derivatives commonly found from acetate biosynthesis in which both the α- and γ-positions bear oxygen substituents, and this caused much confusion in earlier work. The disorder has been straightened out by unambiguous (asymmetric) syntheses and their IR correlations have thereby been put on a sound basis which now allows easy discrimination (α-pyrones ~ 5·8 μ and

γ-pyrones $\sim 6\cdot0\ \mu$). The γ-pyrones are usually more basic and form pyrilium salts with greater facility, whereas the α-pyrones are subject to Diels-Alder reactions, either with maleic anhydride or with themselves. The latter is illustrated in the reaction of paracotoin [17] when heated with maleic anhydride to form [18], or that of phenylcoumalin [17] on KOH fusion, yielding p-phenylbenzoic acid! Both of these simple α-pyrones [17] are constituents of Brazilian rose wood.

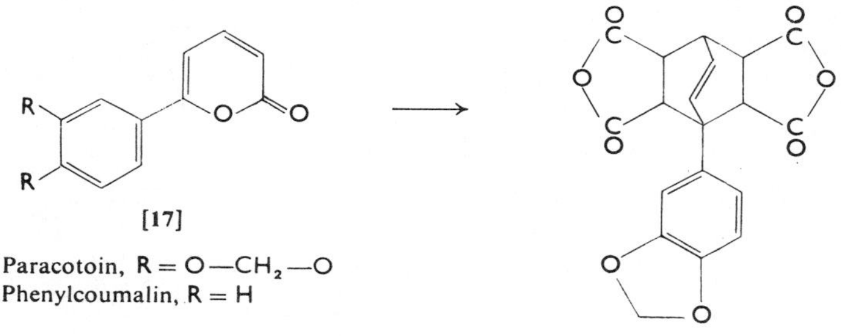

[17]

Paracotoin, R = O—CH$_2$—O
Phenylcoumalin, R = H

[18]

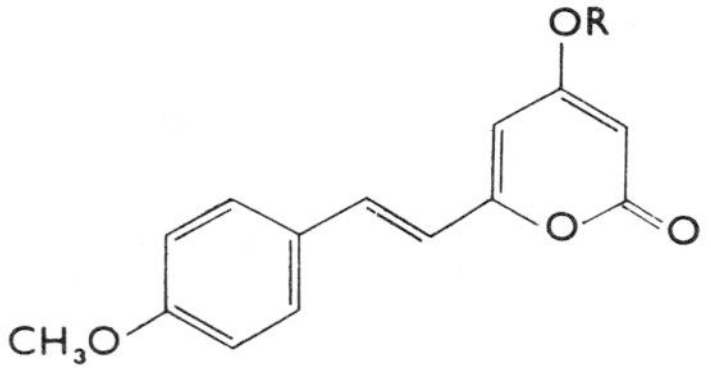

[19]

Yangonin, R = CH$_3$

The difficulties of distinguishing the two systems are obvious in the history of yangonin [19], a pale yellow-green solid with blue fluorescence isolated from the Kawa root, an ingredient central to the preparation of a popular beverage used in the South Seas for ceremonial intoxication. Basic hydrolysis first converts yangonin to ([19], R = H), then opens the ring to afford

$$p\text{-CH}_3\text{O}—\phi—\text{CH}{=}\text{CH}—\text{COCH}_2\text{COCH}_2\text{COOH}.$$

This was degraded further to a ketone by decarboxylation, then by hydrolysis to p-methoxycinnamic acid. The ketone was then synthesized by a Claisen condensation between acetone and ethyl p-methoxycinnamate. However, the hydroxy-pyrone ([19], R = H), yielded both yangonin and an isomer when methylated, and the original workers assumed yangonin was the γ-pyrone because it formed salts with strong acids. Clarification of its structure came only from synthesis: the isomers 4-methoxy-6-methyl-2-pyrone and 2-methoxy-6-methyl-4-pyrone were both prepared and distinguished by the ability of only one (α-) to undergo a Diels-Alder reaction with maleic anhydride (as in [18]) and the greater ease of salt formation (cf. reaction with HCl) of the other (γ-). Yangonin was then prepared unambiguously by aldol condensation, using $Mg(OC_2H_5)_2$, of the α-pyrone with p-methoxy-benzaldehyde. This reaction is often used diagnostically, as in citro-mycin, to indicate the presence of the common 2-methyl group in natural γ-pyrones and chromones.

3-4   Citromycetin

Some of the most intriguing structures among the acetogenins are the mold metabolites, isolated from the culture media of fungi and not infrequently possessing valuable antibiotic qualities. A great many of these were first worked on by Raistrick and his collaborators in an active program which elucidated nearly 200 structures in the period 1930–1950. A number of the more complex of these compounds, left incorrect or unfinished, were taken up by the school of Robertson in Liverpool and there brought to successful conclusions in a series of brilliant and thorough investigations which relied heavily on synthetic confirmation of degradation products.

One of these was citromycetin, an acid with the simple empirical formula $C_{14}H_{10}O_7$ which could be decarboxylated to citromycin, $C_{13}H_{10}O_5$. This product was shown to have two phenolic hydroxyls by methylation to a C_{15} neutral compound, and they were believed to be ortho from the distinctive catechol $FeCl_3$ color. Citromycin shows one $C-CH_3$ in the Kuhn-Roth determination as does a product from $KMnO_4$ oxidation of citromycin dimethyl ether, in which the empirical elements of CH_2 are replaced by CO. The presence of a γ-pyrone was inferred from the formation of a (pyrilium) salt with HCl and the typical 2-methyl of the pyrone was implied by the formation of a styryl derivative (now containing no $C-CH_3$) when citromycin reacted with ϕCHO in base.

Since a 2-methylpyrone requires six carbons and a catechol six more, one more carbon is then CH_2, and all the carbons are accounted for. However, the pyrone and catechol require only four of the five oxygens so that the fifth is assigned to an ether by default. These pieces can be put together in several ways. (■) Formula [20] was the one initially selected for citromycin, although it represents no unique solution.

In the sequel Robertson moved to supply a unique formulation, first via the usual alkaline hydrolysis on the dimethyl ether. From among the products he isolated acetic acid and acetone, as expected for the two β-diketone cleavages of the opened pyrone, as well as the acid ([21], $R_1 = H$, $R_2 = OH$) and the acetophenone ([21], $R_1 = H$, $R_2 = CH_3$), which evidently represented the other products of the ruptured pyrone ring. Put together, these fragments show the direct attachment of pyrone to catechol rings and unveil the position of attachment of the inferred ether link. The $-CH_2-$ carbon is not found in these molecular leftovers but can be placed only as in [22] for citromycin. This was then confirmed by reaction of di-O-methylcitromycin with methyl Grignard followed by acid dehydration to [24] which was synthesized by acidic coupling of acetylacetone with [23]. The position of the carboxyl in citromycetin became clear when the alkaline hydrolysis of methylated citromycetin yielded the phthalic acid ([21], $R_1 = COOH$, $R_2 = OH$) with the original carboxyl intact.

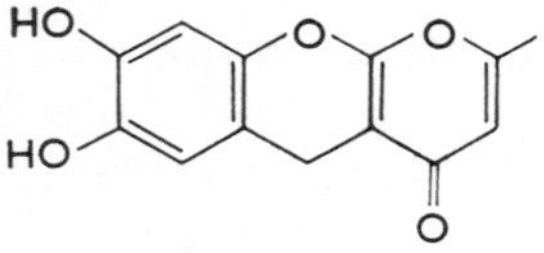

[20]

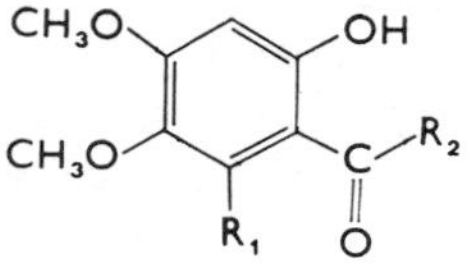

[21]

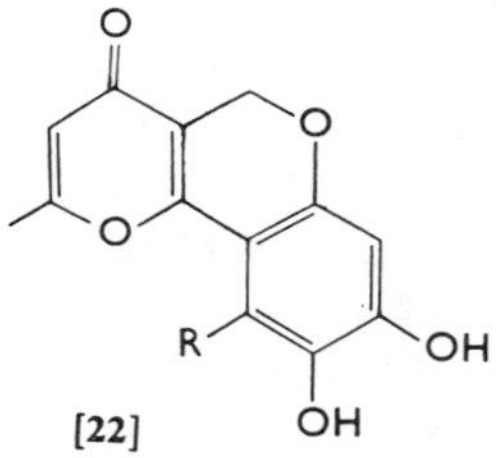

[22]

Citromycetin, R = COOH
Citromycin, R = H

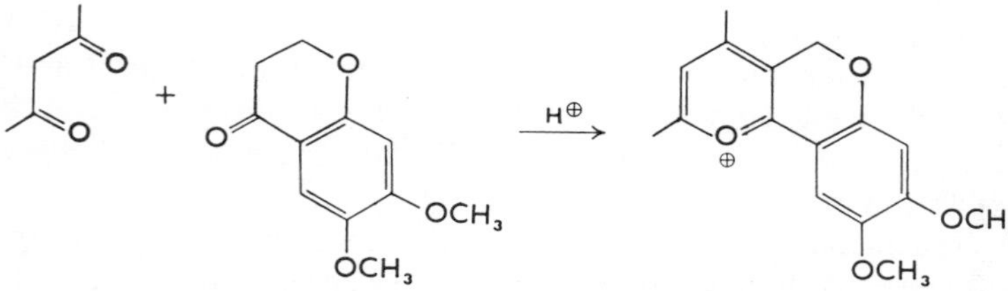

[23]

[24]

3–5   Coumarins, Chromones, and Chromenes

The coumarins and 2-methylchromones often found in nature are only benz-fused pyrones and bear some of the same similarities and confusions common to the simpler heterocycles. The chromones more readily form pyrilium salts with acids and the 2-methyl group commonly reacts with ϕCHO to yield a benzylidene derivative in which the C—CH_3 is lost (Kuhn-Roth). Vigorous alkali treatment will provide salicylic acids from both systems, but only the chromones also give o-hydroxyacetophenones. The coumarins were early distinguished by opening with mild alkali followed by methyl sulfate (since otherwise acidification recloses the coumarin ring) to an o-methoxycinnamic acid. Another early and interesting reaction used in their characterization was bromination followed by alkali, which affords an α-carboxyl-benzofuran.

Biosynthetic isoprenylation of coumarins and chromones is very common, often appearing in the form of simple furans with three isoprenoid carbons missing, as in some of the many constituents of the plants of the *Ammi* species used clinically to combat bronchial asthma spasms: khellin [25], visnagin [25], and bergapten [26]. Khellin was recognized as a 2-methylchromone by its benzylidene derivative and alkaline degradation to the acetophenone [27], from which khellin can be regenerated by Claisen conversion to a β-diketone and acid cyclization. The presence of furans is often recognized by the use of H_2O_2 oxidation which will destroy a vulnerable phenolic ring and produce a furan-2,3-dicarboxylic acid (as from [26], R = H). Both khellin and visnagin are complicated by an isomerization of the furan which occurs on HI demethylation, yielding [28] from the visnagin. Curiously enough, similar isomerizations can even occur in alkali, for while treatment of bergaptol [26] with CH_2N_2 yields bergapten, $(CH_3)_2SO_4$ in alkali produces isobergapten [29].

Isoprenoid substitution is most commonly found in its simplest form, as the γ,γ-dimethylallyl substituent either on enolic (or phenolic) oxygen, or its adjacent carbon (Chapter 2; also, mangostin, p. 78). It is usually recognized chemically by ozonolysis to acetone, and/or by hydrogenation to an inert isoamyl substituent, later characterized by synthesis of a suitable degradation product, and physically by its distinctive usually unsplit NMR spectrum of six allylic methyl protons and two methylene protons at lower field. A number of isoprenoid substituents are cyclized, however, forming 2,2-dimethyl-

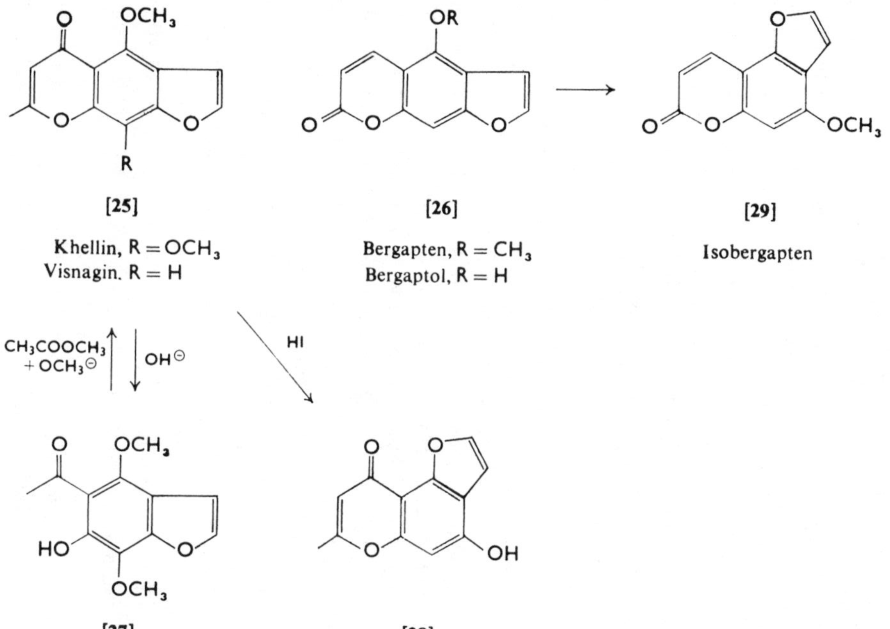

[29]
Isobergapten

[26]
Bergapten, R = CH₃
Bergaptol, R = H

[28]

[25]
Khellin, R = OCH₃
Visnagin, R = H

[27]

chromene rings in many cases. This is a common circumstance not only in the coumarins, chromones, and flavonoid molecules, but also in the biogenetically related and very common natural acetophenones, of which evodionol may serve as an illustrative sample.

A phenol bearing one methoxyl and a methyl ketone (via ketone derivatives and a benzylidene one), evodionol ($C_{14}H_{16}O_4$) absorbs one mole of hydrogen catalytically to yield a phenolic dihydro compound. From this evidence and the assumption (by default) that the fourth, unreactive oxygen is an ether, evodionol must have two rings: one the phenol, the other a ring that is opened by cold $KMnO_4$ oxidation on evodionol methyl ether to yield a dicarboxylic acid, evodionic acid ($C_{15}H_{18}O_8$). This must correspond to cleavage of $-CH=CH-$ in a ring by virtue of the empirical composition change. Evodionic acid on treatment with hot mineral acids produces 4-hydroxy-2,6-dimethoxy-acetophenone and α-hydroxyisobutyric acid. When this product aceto-phenone was then treated with methyl α-bromoisobutyrate in alkali, a monoacid, identical with one produced by hot acidic decarboxylation of evodionic acid, resulted. These reactions, typical of the degradation of dimethylchromenes generally, in this case establish all but the choice of which phenolic hydroxyl is methylated in evodionol [30]. (■) One more feature of chemical interest is the dimerization of many natural dimethylchromenes with concentrated acid, a mechanistically interesting example being the formation of [32] from lapachenole [31].

3–6   Rotenoids

Interesting features of many of the foregoing systems are found together in the natural rotenoids, which are biogenetically related to the isoflavones. These compounds are widely used commercially as insecticides since they are not toxic to mammals. As they also paralyze fish without rendering them toxic, rotenoid-bearing roots are used in Africa for lazy fishing.

The main requirements for a structural analysis of rotenone ($C_{23}H_{22}O_6$)—although they represent an oversimplification of its actual history—include first the H_2O_2 oxidation under alkaline conditions to dehydronetoric acid, the structure of which [33] was confirmed by Robertson's synthesis. (Netoric is an anagram of rotenic, an occasional desperate practice in natural products nomen-clature.) Secondly, vigorous KOH treatment of rotenone affords tubaic acid, recognized as a salicylic acid by its properties. Catalytic hydrogenation of this produces optically active dihydrotubaic acid and, more slowly, optically inactive tetrahydrotubaic acid, which can be decarboxylated to 2-isoamylresorcinol. Tubaic acid is also isomerized by strong acids to optically inactive isotubaic acid.

From these results we are able clearly to deduce the structure [34] for tubaic acid. This was confirmed by Reichstein's ingenious synthesis of isotubaic acid [35], which is a rare example in that it builds a benzene ring onto a furan instead of the reverse: a Stobbe condensation, [36] → [37], provides the requisite carbons, which cyclize (and

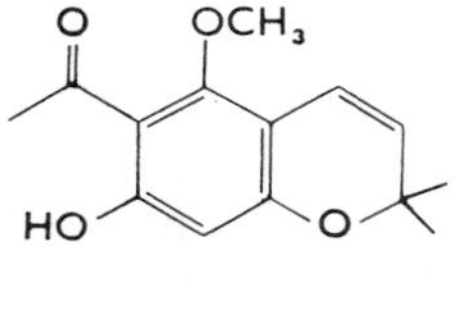

[30]

Evodionol

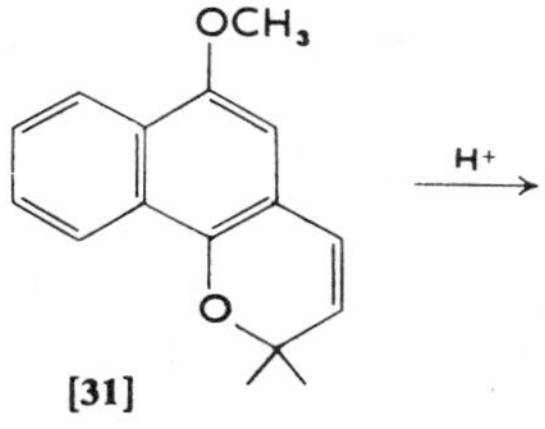

[31]

Lapachenole

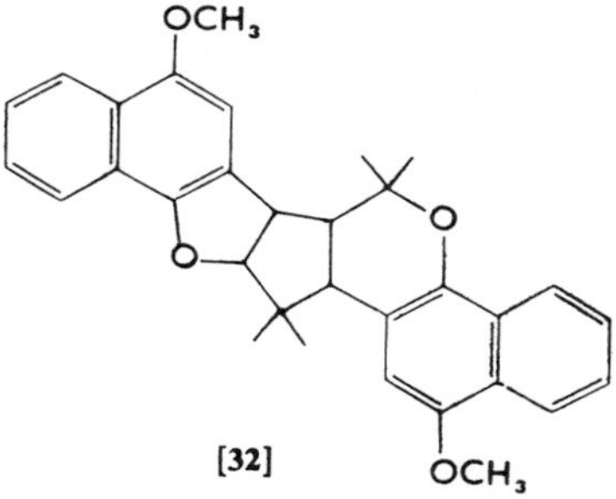

H^+ →

[32]

decarboxylate) on heating; the salicylic acid carboxyl is introduced afterwards.

These two fragments, dehydronetoric and tubaic acids, quite evidently represent two separate moieties of the molecule and together contain twenty-four carbons, so that we assume only one atom of rotenone (C_{23}) is common to both and that one will most likely be the carboxyl each possesses. If we thus combine the fragments through a common ketone, the molecule obtained would be an isomer of rotenone which needs only a Michael cyclization to achieve the correct formula [38].

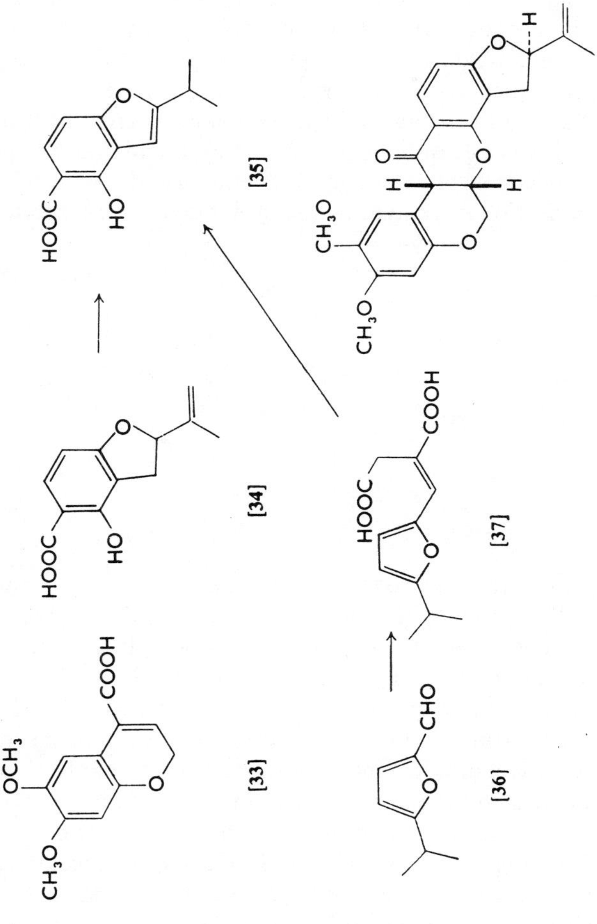

[33]

[34]

[35]

[36]

[37]

[38]
Rotenone

The three asymmetric centers indicated in [38] may be readily destroyed in two steps: (1) dehydrogenation to dehydrorotenone, a chromone with physical properties akin to isoflavones; (2) strong acid isomerization of the isopropenyl side chain to an isopropyl-furan as in isotubaic acid. Oxidation of CH_2 to CO in dehydrorotenone can be effected (as in citromycin) and the product, rotenonone [39], is sensitive to alkali, yielding an acid which when heated affords the lactone [40] in a nice display of disguised carbonyl reactivity. The central ring fusion in rotenone as in all rotenoids is *cis*. Its absolute stereochemistry was elegantly demonstrated by Büchi and Crombie, as follows. Conversion of rotenone to its enol acetate was achieved without epimerization since hydrolysis regenerated natural rotenone.

[39]　　　　　　　　　　　　　　[40]

The side chain was then hydrogenated and the enol acetate ozonized exhaustively. The products included D-(−)-glyceric acid from the center of the molecule and (+)-3-hydroxy-4-methylpentanoic acid from the side chain, thus establishing at one stroke the absolute configurations at both of those asymmetric centers as S− and R−, respectively, as illustrated in [38].[2] Since all rotenoids exhibit a similar positive Cotton effect in ORD measurements, they probably all have the same absolute stereochemistry at the central junction.

The introduction of NMR and mass spectrometry have been very useful in the flavonoid-rotenoid field. In the mass spectrometer, fragmentation of rotenoids conveniently occurs in the middle of the molecule to give the two important fragments [42] and [43] so that the nature of their substituents is usually readily deduced from their masses. The wide range of chemical shifts in the NMR, on the other

---

[2] Parenthetically, we may note Barton's similar and very elegant determination of the absolute stereochemistry of terrein [41] by ozonolysis of its diacetate directly to the known (+)-tartaric acid diacetate.

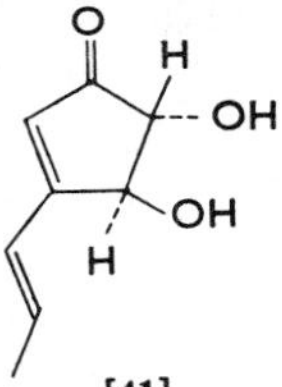

[41]
Terrein

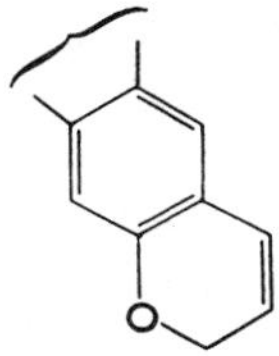

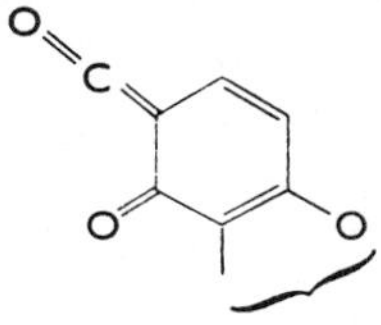

[42]

+

[43]

hand, allows the identification of every proton and their orientations in many cases from the splittings. Combined with IR and UV spectrometry, these tools allowed Ollis and his collaborators to assign structures rapidly to a variety of constituents of *Mundulea sericea*, a plant reportedly used by African natives to disperse alligators from streams during cattle-crossings. Very small quantities of these compounds sufficed for an elucidation, and indeed even elementary analyses were unnecessary in some cases as the physical methods led directly to structures which could be more readily confirmed by synthesis!

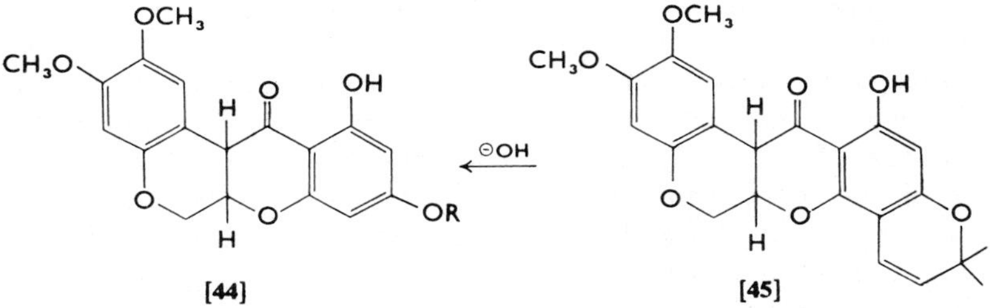

[44]

Sermundone, R = CH$_3$
Apotoxicarol, R = H

[45]

α-Toxicarol

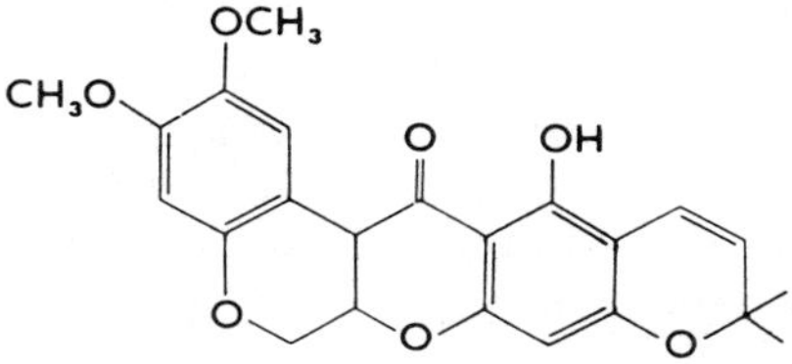

[46]
β-Toxicarol

Such was the case with sermundone [44], in which the mass spectrum showed the characteristic two fragments of masses 192 [42] and 166 [43] as well as the molecular weight (358); NMR revealed the three methoxyls and nine other protons, those on the phloroglucinol showing (characteristically) very high field values and meta-orientation, the others para-orientation; and the UV and IR confirmed the o-hydroxy-ketone. With the formula assigned, confirmation lay simply in mild methylation of apotoxicarol [44] of known structure, it being common to find the ortho-hydroxy difficult to methylate because of hydrogen-bonding stabilization.

Finally, the formation of apotoxicarol and acetone from alkaline hydrolysis of the natural rotenoid, α-toxicarol [45], provides a nice mechanistic puzzle; a further characteristic of toxicarol is the alkali-induced conversion of the natural optically active rotenoid to a mixture of *racemic* α- and β-[46] toxicarols.

3–7   Xanthones: Mangostin

The latex of the Malayan mangosteen tree yields the yellow crystalline pigment, mangostin ($C_{24}H_{26}O_6$), regarded as a xanthone from its characteristic four-banded UV absorption, which is also shown by the tetrahydro-derivative. Of the three phenolic groups two are easily methylated and the third, by spectral evidence also, is placed in the hydrogen-bonded 1-position. As mangostin also possesses a methoxyl, all six oxygens are thereby recognized. Oxidation of the tetrahydro compound yielded *more* than one mole of isocaproic acid, thus implying *two* isoamyl side chains and accounting for all the carbons. Alkali fusion had yielded isoamyl alcohol, isovaleric acid, phloroglucinol, and a phenol, $C_{12}H_{16}O_3$, containing the one original methoxyl of mangostin. The dimethyl ether of this phenol afforded 2,3,5-trimethoxybenzoic acid on oxidation while the phenol itself was not oxidizable to a quinone. Yates showed that ozonolysis of mangostin produced a C_{18} dialdehyde in which the CHO groups were unconjugated (by IR: 5·80 μ) and no C—CH_3 groups remained (by NMR). Finally HI demethylation yielded two isomeric and very similar desmethyl-mangostins which were found to be saturated to catalytic hydrogenation. (■) These facts allowed Yates to write the structure [47] for mangostin and rationalize the mechanism for alkaline release of isoamyl alcohol and isovaleric acid, the last step of which is a Cannizarro disproportionation of isovaleraldehyde.

3-8   Griseofulvin; Curvularin

Strains of the richly productive *Penicillium* fungi elaborate a crystalline metabolite called griseofulvin ($C_{17}H_{17}O_6Cl$), originally studied by Raistrick and more recently in great detail by several pharmaceutical laboratories. Griseofulvin is effective at arresting fungal growths in plants and animals while remaining harmless to the host; its utility arises because it can be administered orally to animals, and because it is readily absorbed by plant roots and subsequently suffused throughout the plant. Commercial fermentations now produce griseofulvin in yields of up to 5 grams per liter of culture medium.

Griseofulvin is an unsaturated ketone (UV) with three methoxyls, one of which is easily hydrolyzed by dilute acid with attendant formation of a β-diketone. The griseofulvin molecule is conveniently ruptured into two roughly equal halves by the action of methoxide, which yields the salicylic acid [48] (also obtained by $KMnO_4$ oxidation) and orcinol methyl ether (3-hydroxy-5-methoxy-toluene). The latter product evidently arises from the same portion of the molecule which produces 2-methyl-6-methoxy-*p*-benzoquinone on CrO_3 oxidation. The products [48] and the orcinol ether contain between them all the atoms of griseofulvin. Raistrick was led astray in his deductions by the formation of CO_2 and [49] in dilute alkali, or the corresponding phenolic dibenzofuran ([49] with 2H less) which is produced instead when air is present. The correct structure for griseofulvin [50] may be deduced from the milder $KMnO_4$ product, a keto-hydroxy-acid ($C_{14}H_{15}O_7Cl$) which readily forms a keto-γ-lactone (IR evidence) and is cleaved quantitatively to (+)-methylsuccinic acid and [48] by the action of HIO_4. (■)

An interesting stereochemical change is produced by hot dilute NaOMe which affords an equilibrium mixture of griseofulvin and an epimer which yields the same (+)-methylsuccinic acid on oxidation so that the change must involve epimerization of the quaternary center by a base-catalyzed mechanism.

Several syntheses of griseofulvin have been reported, that of Scott simulating the proposed biogenesis by ferricyanide oxidation of the benzophenone [51][3] to dehydro-griseofulvin, the cyclohexadienone corresponding to griseofulvin and convertible to the natural

---

[3] This benzophenone has been isolated as a natural product from the same mold that produces griseofulvin.

compound by hydrogenation, although this step normally is com-
plicated by extensive hydrogenolysis of the ether link at the quaternary carbon, leading back to [51]. Stork's ingenious synthesis is

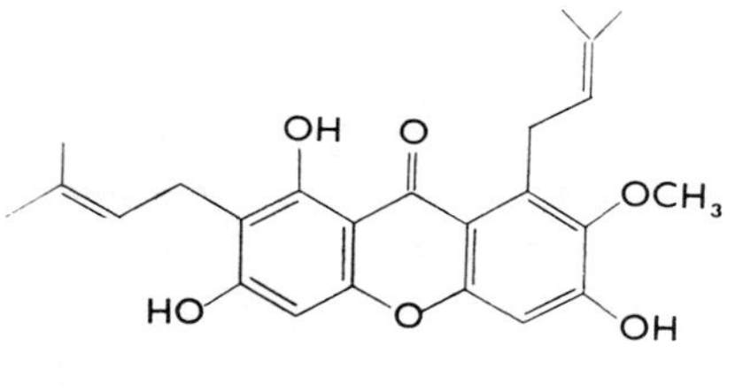

[47]
Mangostin

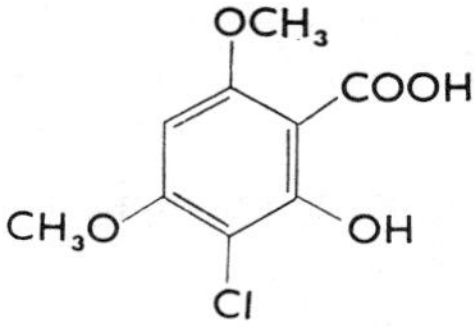

[48]

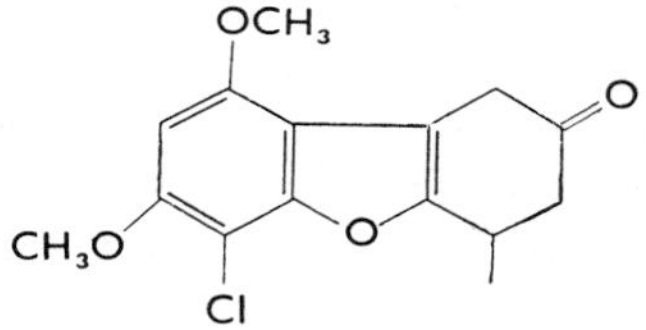

[49]

illuminating in showing the stimulus the synthesis of natural products
provides to the development of new reactions, for Stork envisioned
here a double Michael addition of the anion of the ketone [52] on the

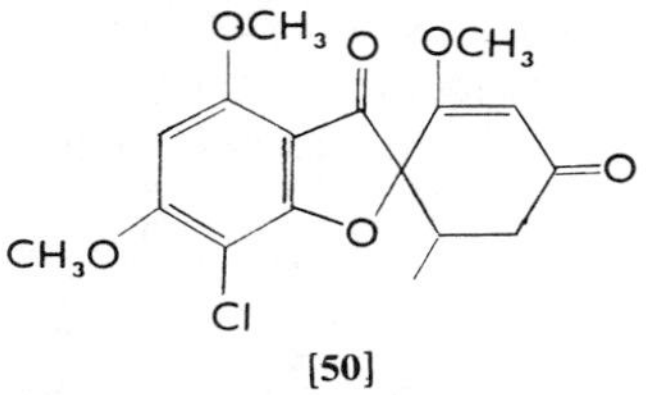

[50]
Griseofulvin

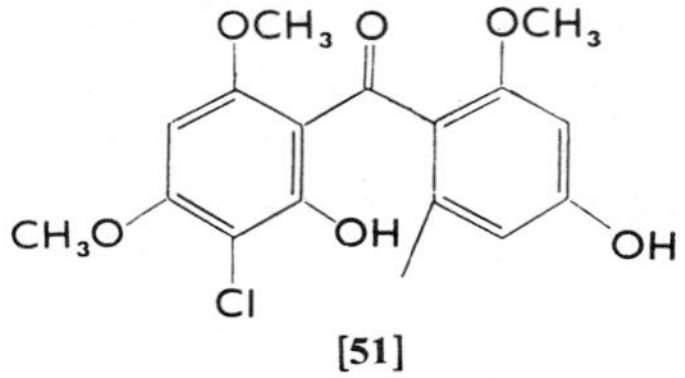

[51]

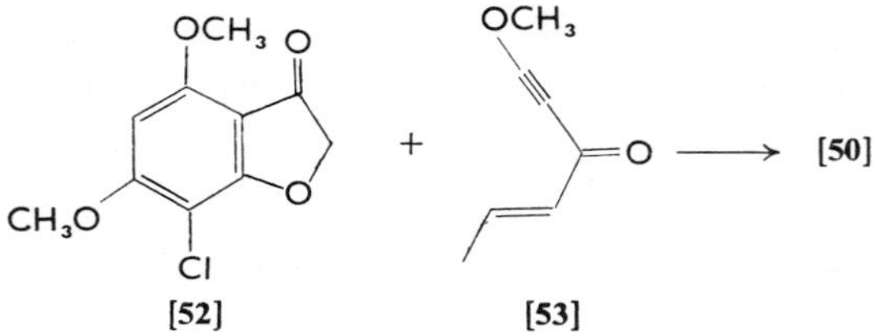

[52]   +   [53]   ⟶   [50]

CHART 3 ▪

Degradation of Radioactive Griseofulvin

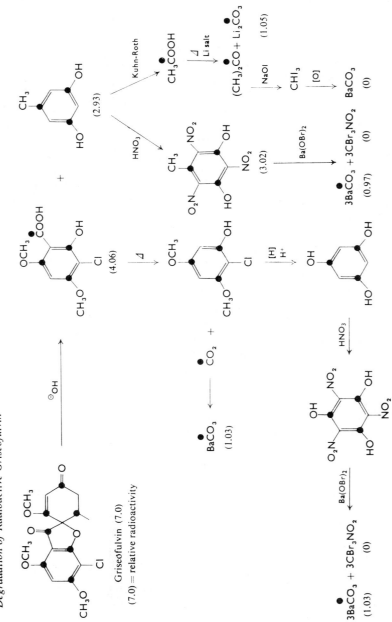

Griseofulvin (7.0)

(7.0) = relative radioactivity

doubly unsaturated ketone [53]. Not only was the double Michael addition without previous precedent, but alkoxyacetylenic ketones were also unknown. First developing a successful method for syn. thesis of the latter, he then showed that the Michael addition in fact proceeds smoothly and yields griseofulvin, exclusively, with the correct stereochemistry, in one step!

An impressive body of tracer studies has been carried out by Birch and his collaborators to confirm the biosynthetic routes to aceto-genins outlined in Chapter 2. Their studies on the biosynthesis of griseofulvin are fairly representative of the procedures used in many of these natural polyphenols. The mold is grown on a medium containing radioactive acetate, in this case sodium $1\text{-}C^{14}$-acetate, and the metabolite is harvested and purified. The skein of degradations carried out to isolate various carbons is outlined in Chart 3; the carbons presumed (and proved) labeled are shown as dots, and the observed relative reactivity of each molecular species is shown beneath each dot (relative to initial griseofulvin taken for convenience as 7·0). The most common reactions used in this tracer work and exemplified in the chart are the Kuhn-Roth oxidation and the bromopicrin degradation, so well suited for the resorcinols and phloroglucinols encountered among acetogenin structures.

Birch has carried this biogenetic tracer work into the field of structure elucidation itself in the case of the mold metabolite, curvularin, a $C_{16}H_{20}O_5$ phenolic compound which naturally suggested a bio-origin in eight acetate units. The compound had a lactone grouping and one C—CH_3, and it turned out that if the mold were grown on $1\text{-}C^{14}$-acetate, the acetic acid produced in the Kuhn-Roth determination had just one-eighth the activity of the parent curvularin, as anticipated biogenetically. HBr cleavage of the molecule, recognized spectroscopically as a simple 4-keto-resorcinol, yielded two C_8 fragments of equal radioactivity. (■) These facts lead directly to the structure [54] if the biosynthetic considerations implied by the tracers here are utilized. The dimethyl ether of curvularin on treatment with alkali isomerizes to a naphthol.

3–9 Orsellinic Acid

Most of the compounds treated so far in this chapter are biogenetically of the phloracetophenone family (Chapter 2; p. 24). The other mode of polyacetyl chain cyclization, exemplified most simply in the widespread orsellinic acid [55], is very common among the lower organisms, lichens, and fungi. In fact, it was Collie's synthesis of orsellinic acid by alkaline treatment of dehydracetic acid (itself obtained by pyrolytic dimerization of ethyl acetoacetate and thus

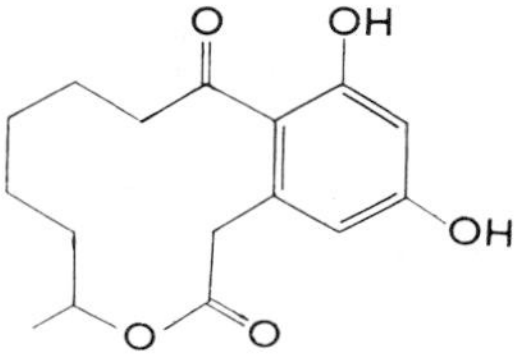

[54]
Curvularin

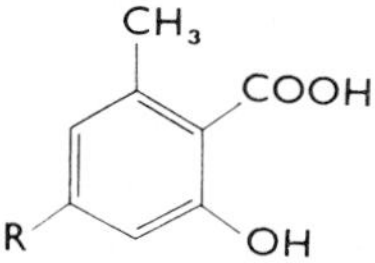

[55]
Orsellinic Acid, R = OH

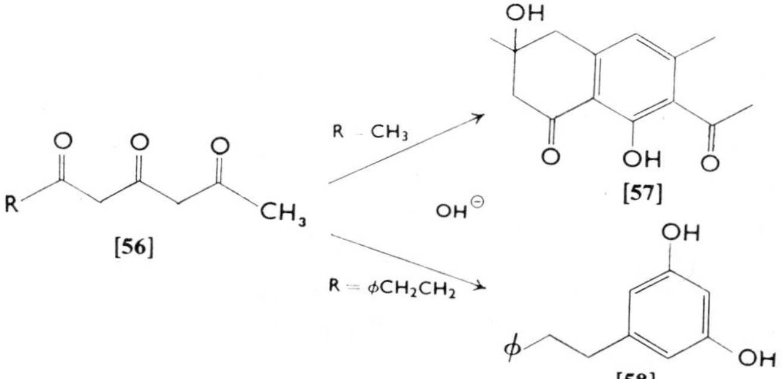

[56]

R CH₃

OH⁻

[57]

R = φCH₂CH₂

φ

[58]

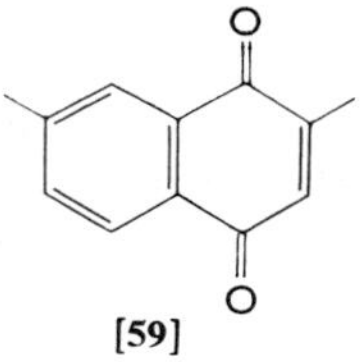

[59]
Chimaphilline

equivalent to "four acetate units") that led to his first hazy speculations on linear polyacetate as a biogenetic source in 1893. In his extensive explorations of the acetate hypothesis, Birch has extended Collie's laboratory models for biogenesis by synthesis of the poly-β-ketones [56] and their conversion in alkali into [57], reminiscent of the natural naphthaquinone, chimaphilline [59], and into the dihydro derivative [58] of pinosylvin, one of the insecticidal stilbenes of pine heartwood.

Orsellinic acid has been one of the most intensively studied compounds in the probing of acetate biosynthesis. Swedish chemists, feeding O^{18}-acetate, showed that the organism used it directly to synthesize orsellinic acid retaining all the O^{18}. Others showed that the degradation of the related metabolite 6-methylsalicylic acid [55] (R = H) by a Kuhn-Roth procedure (producing $CH_3COOH + CO_2$) yielded all its radioactivity in the CO_2 when the organism was fed C^{14}-malonate, but 25 per cent in the CH_3COOH and the rest in CO_2 when fed labeled acetate. Finally, Lynen has isolated an enzyme from *Penicilium* fungi which stoichiometrically catalyzes the reaction of one mole of acetyl coenzyme A and three moles of malonyl coenzyme A to produce 6-methylsalicylic acid.

3–10   Depsides and Depsidones

As the mild, moist climate of Japan is admirably suited to the luxuriant growth of lichens, it comes as no surprise that their chemistry has been extensively studied there, chiefly by Asahina in the 1930's. It is not uncommon to find partial evaporation of a filtered chloroform extract of dried lichens producing a crop of lovely golden crystals. The major group of lichen substances is that of depsides and depsidones (Chapter 2—Chart 5), the former simply arising by formation of an ester link between two molecules of a substituted orsellinic acid, the latter by oxidative coupling of the two aromatic rings of a depside to form an ether bridge.

The diphenyl ether link in the depsidones is generally impervious to acidic or basic solvolytic disruption so that the ordinary analytical reactions of substituents on the two aromatic rings may be carried out with impunity in the dimeric series. Oxidation, on the other hand, serves conveniently to cleave the depsidone in two since the depsidone ring B is almost invariably a substituted hydroxy-hydroquinone which appears after the oxidation as a hydroxyquinone (HIO_4 is the reagent of choice although several others have been used). Physodic acid [60] may be taken as representative and also includes the biogenetically reasonable (and fairly common) β-keto odd-membered side chain. Saponification of the lactone results in isophysodic acid [61] which, after methylation and treatment with hot alkali, suffers hydrolysis, decarboxylation, and β-diketone fission to furnish the relatively bare

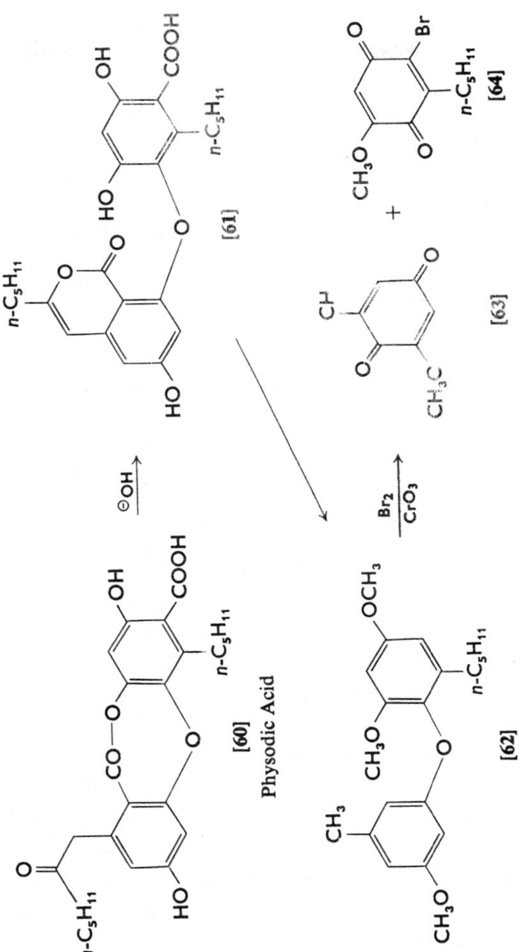
[60]
Physodic Acid
[61]
[62]
[63]
[64]
$^{\ominus}OH$
Br_2
CrO_3

diphenyl ether [62]. Oxidations on this ether now yielded the quinones [63] and [64] from the two rings, and [62] was synthesized with an Ullman coupling.

3–11 Usnic Acid

The most common lichen substance is probably usnic acid, a yellow acidic enol comprising up to 20 per cent of the dry weight of some lichens (although its biological function, if any, has not been discerned!). It was first isolated and examined in the 1840's but abandoned when it became clear its complexity was beyond the reach of the then current understanding. It was finally successfully attacked in the 1930's, primarily by Asahina and by Robertson.

In polyphenols of roughly this molecular size, degradations (usually with alkali) commonly proceed in one of two ways: either a weak link snaps in the center of the molecule leaving two roughly equal-weight products from the two halves, as has been the case in many families in this chapter (e.g., the flavones); or the degradation leads to an attrition, by removal of small fragments, which increases with increasing vigor of the conditions until a resistant nub of the original molecule remains. The latter case is represented in usnic acid and the common analytical history of such cases is one of identifying the "nub" and synthesizing this, then working back up the degradative ladder, often via synthesis, formulating increasingly larger molecules until the original natural product is reached.

The mildest degradation of usnic acid ($C_{18}H_{16}O_7$) occurs when it loses only CO_2 (to give decarbousnic acid) on heating in alcohol. Successively more vigorous alkaline attacks lead to a succession of products culminating in the base-stable pyrousnic acid ($C_{12}H_{12}O_5$) which, however, loses CO_2 on heating to give usneol, $C_{11}H_{12}O_3$, a dibasic phenol with no carbonyl reactivity. Ozonolysis of usneol yields a compound $C_{11}H_{12}O_5$, empirically implying the cleavage of a simple double bond in a ring. This was then easily shown to be an acetate of C-methylphloroacetophenone by further hydrolysis, and it can be only [65] (or its isomer with CH_3 at →) if it is to be derived by an ozonolytic cleavage of a ring. This proved the presence of a benzofuran system in usneol ([66], R = H), the ether of which was then synthesized for verification by condensing 3-chlorobutanone with 4-hydroxy-2,6-dimethoxytoluene, and cyclizing in acid.

The next step involved locating the carboxyl of pyrousnic acid:

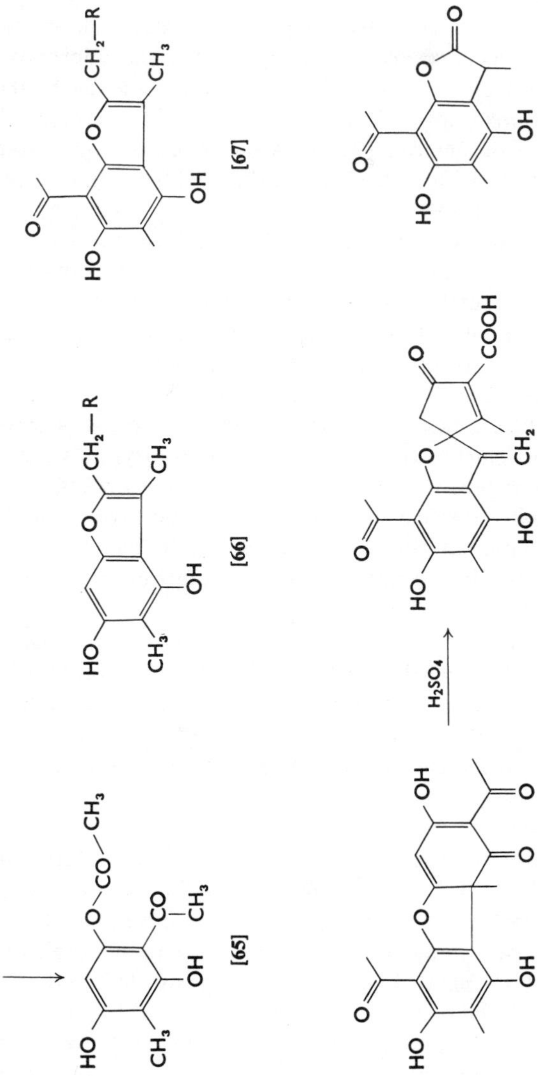

[65] [66] [67]

[68] Usnic Acid

[69] [70]

H_2O_2 oxidized this acid to 3-methylfuran-2,4,5-tricarboxylic acid so that its constitution was assigned as [66] (R = COOH) and confirmed by synthesis. Farther back up the degradation cascade lay usnetic acid, empirically an acetyl derivative of pyrousnic acid and converted to that compound by heating with 50 per cent aqueous KOH. These strong conditions imply a C-acetyl group and, mechanistically, there is only one reasonable place for it; so usnetic acid was presumed to be [67] (R = COOH) and its ether synthesized from that of pyrousnic acid with Ac_2O/BF_3.

Now decarbousnic acid yielded acetone and usnetic acid on alkaline hydrolysis, accounting for all its carbons and implicating a β-diketone chain, which was further supported by its reaction with hydroxylamine to give an isoxazole. Thus decarbousnic acid is [67] (R = CO-CH$_2$-COCH$_3$); but the structure [68] deduced from it for usnic acid, having but one more carbon, was a cause for much debate, which will come as no surprise when some of the other reactions of usnic acid are outlined. However, the clarification of polyacetate biosynthesis which has developed since [68] was formulated allows us to recognize in usnic acid two equal phloracetophenone halves oxidatively coupled, thus offering dramatic support to the structure. Its elegant synthesis by Barton, who duplicated this coupling with ferricyanide, was the first of such oxidative biosynthetic emulations.

If the elucidation of usnic acid did not run a simple course, this may be understood by considering the curious observation that it is racemized in boiling toluene and that its conversion to usnolic acid [69] by concentrated H_2SO_4 was bound to mislead investigators! Finally, pyrolysis of dihydrousnic acid diacetate with $CaCl_2$ yields [70].

3–12   Natural Quinones

Lichens and fungi also produce some bronze-colored pigments which all yield p-terphenyl on zinc dust distillation. If we accept that skeleton for polyporic acid, $C_{18}H_{12}O_4$, we have only to place the oxygens on it, and since oxidation yields more than a mole of benzoic acid (■), both terminal rings must be unsubstituted, and all the oxygens must be placed on the central ring. Hence polyporic acid can only be [71]. The pigment kindly offers a mechanistic puzzle in its conversion to α-benzylcinnamic acid and oxalic acid when boiled with aqueous KOH. Furthermore, Pb(OAc)$_4$ oxidation of polyporic acid leads to pulvic anhydride [72], itself a natural fungal pigment and supposed to arise in nature by such an oxidation.

The quinones in nature form a massive collection of natural coloring matters with representatives in all three classes: benzoquinones (cf. polyporic acid), naphthoquinones, and anthraquinones, the latter being the most widespread (there are nearly a hundred known examples). The anthraquinones, though numerous, usually conform fairly closely to the biogenetic pattern with a β-methyl and at least one hydrogen-bonded α-hydroxyl among several phenolic groups. Thus, their recognition by absorption spectroscopy (and spectral shifts with bases or chelating agents) is usually easy and goes far toward defining even the phenolic orientations. In fact, this much was clear long before spectrophotometers were developed, since the absorptions and shifts involved are mostly in the visible range.

Alizarin [73] was isolated in 1826 as the colored constituent of madder root, the most ancient and important of natural dyestuffs. It was the subject of a classic investigation by Graebe and Liebermann, who recognized its skeleton from the formation of anthracene on zinc dust distillation and its nature as a dihydroxy-anthraquinone from certain synthetic quinone studies concurrently being pursued. They promptly synthesized it in two ways (cf. KOH fusion of anthraquinone-β-sulfonic acid) in 1868–69, the first synthesis of a natural product (of more than two carbons), and immediately patented it. Synthetic alizarin was on the market by 1870 and commercial cultivation of the madder plant (*Rubia tinctorum*) rapidly declined.

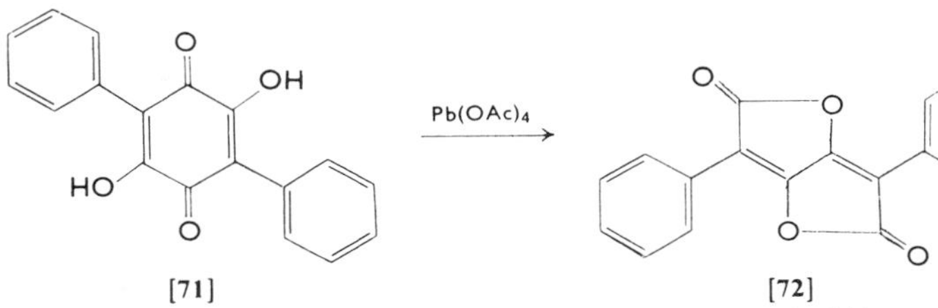

[71]
Polyporic Acid

[72]
Pulvic Anhydride

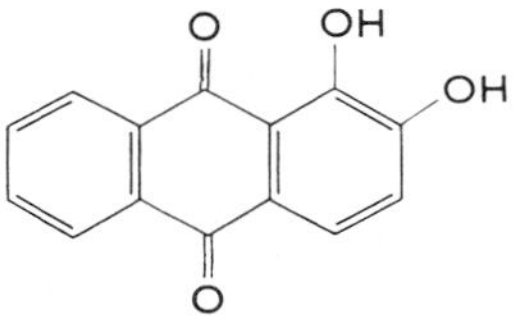

[73]
Alizarin

Many of the natural naphthalenes are naphthoquinones and include the metabolically important electron-transfer agents, vitamins K [74]. The biologically analogous coenzymes Q are benzoquinones [75]. A representative and interesting naphthoquinone, isolated from various tropical woods in the last century, is the yellow dye lapachol, $C_{15}H_{14}O_3$. As is common with quinones, the quinonoid nature of lapachol follows from its reversible reduction (with loss of color) and is confirmed by monoacetylation of lapachol and triacetylation of its reduced form; the other oxygen atom is simultaneously recognized as an acidic hydroxyl. Zinc dust distillation to naphthalene and peroxide oxidation to phthalic acid and acetone reveal a naphthalenic skeleton with all the substituents on one ring and leave a five-carbon side-chain, probably terminating in $C=C(CH_3)_2$, to be formulated. This chain can only be placed at the 3-position of a 2-hydroxyl-1,4-naphthoquinone (or its o-quinone tautomer). Lapachol [76] forms two neutral quinonoid isomers $(C_{15}H_{14}O_3)$ on acid treatment and undergoes a much more subtle transformation to [77] (apparently an abduction of one methylene group!) with dilute $KMnO_4$ (or H_2O_2) in alkali.

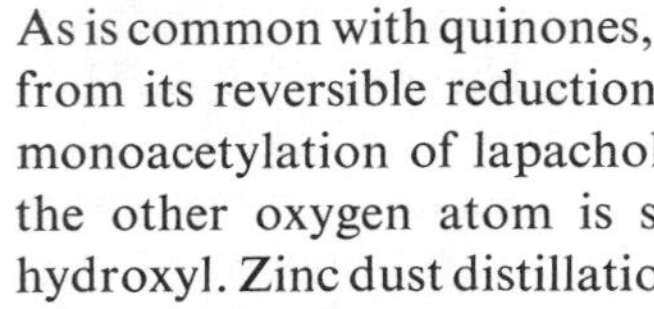

[74]
Vitamin K$_2$

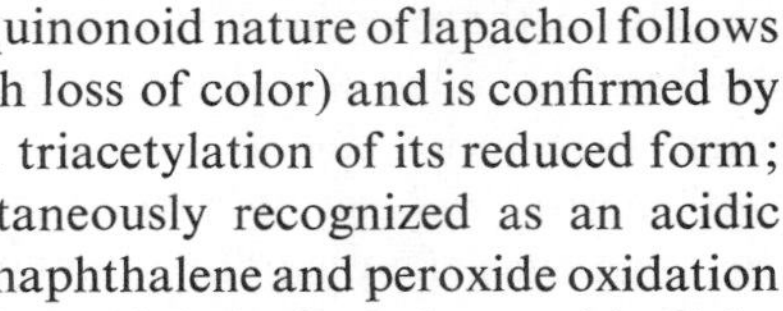

[75]
Coenzymes Q

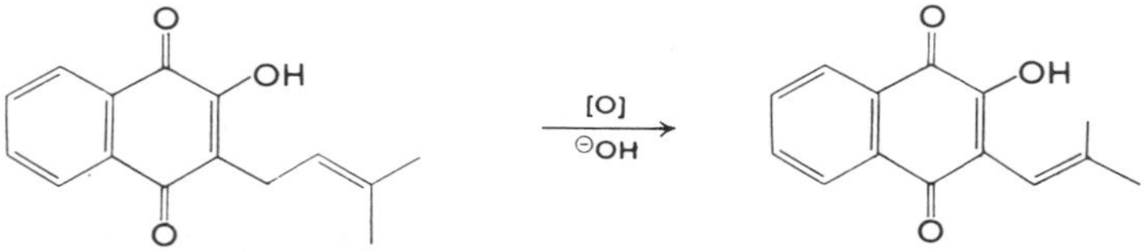

[76]
Lapachol

[77]

3–13   Citrinin

Not all natural quinonoids are so simple, as the interesting reactions of citrinin can attest. This yellow acidic mold metabolite $(C_{13}H_{14}O_5)$

yields a colorless dihydro derivative which is phenolic and can be reoxidized to citrinin, thus suggesting a quinonoid nature. On acidic or basic hydrolysis citrinin loses CO_2 and HCOOH to form a $C_{11}H_{16}O_3$ alkyl-resorcinol, and the early formulation of citrinin as a quinone-methide (but the wrong structure) by Raistrick and Robinson in 1931 led to considerable interest in its chemistry. The resorcinol loses its optical activity during strong alkali degradation to [78] and acetic acid (concomitant aerial oxidation also occurs). Robertson established the identity of [78] by a synthesis in which an aldehyde group is introduced into 5-ethylresorcinol (via $Zn(CN)_2/HCl$) and the isomer produced identified by the fact that it forms two (not one) monomethyl ethers; Wolff-Kishner reduction then yielded [78]. He then proceeded to synthesize both possible diastereomers of the dimethyl ether of the C_{11} resorcinol [79]. The natural optically active [79] from citrinin is interesting in that it is easily racemized (*not* just epimerized) in acid despite the presence of *two* asymmetric centers. Cram's later study of this natural product led to his formulation (and subsequent elaboration) of the phenonium ion participation mechanism (cf. intermediate [80]) which both explained this unusual property and demonstrated the stereochemistry to be that shown in [79].

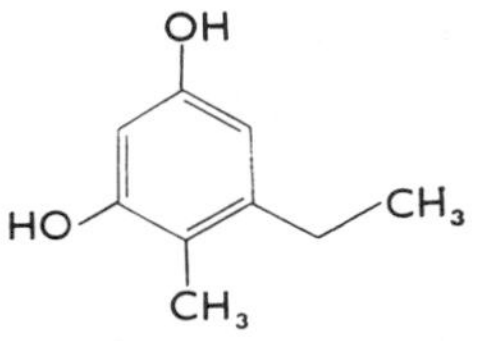

[78]

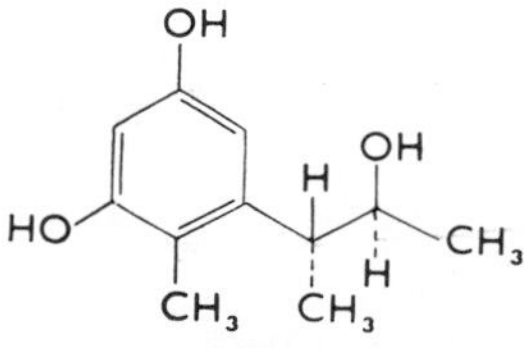

[79]

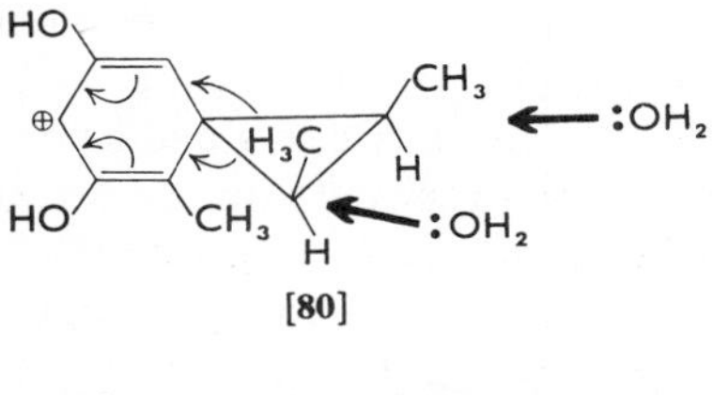

[80]

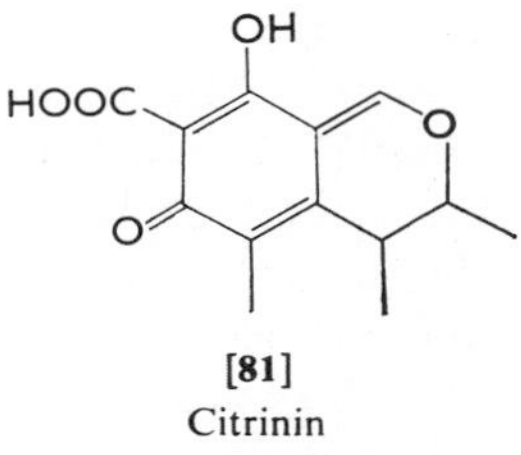

[81]
Citrinin

In order to arrive at a formula for citrinin from the C_{11} resorcinol we must reintroduce the CO_2 and HCOOH lost on hydrolysis in such a way as to regenerate its quinonoid properties. Robertson also achieved this in the laboratory by Kolbe carboxylation plus a Gattermann aldehyde synthesis, which afforded citrinin [81]. The interesting reactivity of this molecule is implied not only by its hydrolysis but also by its formation of a simple *colorless* phenolic *adduct* (no loss of H_2O) with phenylhydrazine, and by its methylation with $(CH_3)_2SO_4$ and base to the neutral $C_{15}H_{18}O_5$, which gives diethyl ketone on basic hydrolysis and a C-methyl derivative of [79] with aqueous acid.

3–14   Tropolones; Patulin

There are several mold metabolites with very simple empirical formulas which nevertheless caused considerable agony in their

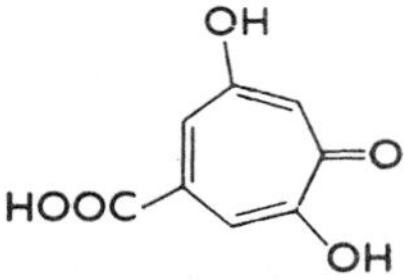

[82]
Stipitatic Acid

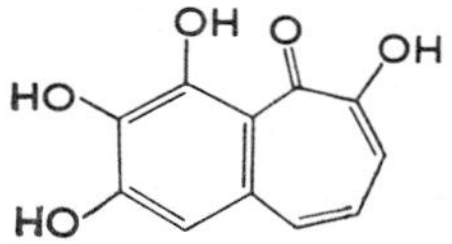

[83]
Purpurogallin

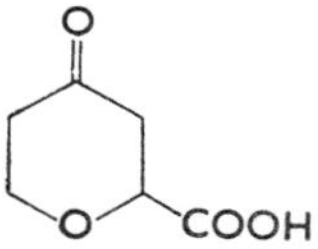

[84]

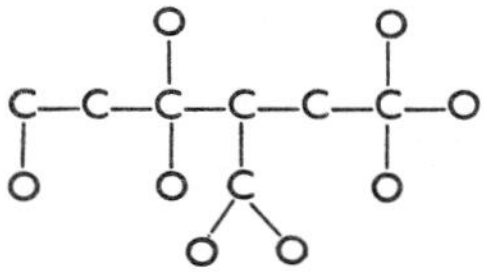

[85]

elucidation. Perhaps the most notorious are a group of related acids represented by stipitatic acid, $C_8H_6O_5$, which gives a $FeCl_3$ test and a trimethyl derivative with CH_2N_2. Its conversion via KOH fusion to the *isomeric* 5-hydroxybenzene-1,3-dicarboxylic acid put a premium on unusual formulations which completely stymied the original workers. Dewar's brilliant vision of the tropolone ring and its chemistry in 1948 arose directly from this problem and led to his formulation of stipitatic acid as [82]. Some related and fascinating reactions are found in Haworth's demonstration that the natural benztropolone pigment purpurogallin [83] may be formed in the laboratory by $O_2/$ $OH^{\ominus}$ oxidation of pyrogallol and is subsequently converted to β-methyltropolone by further action of air and alkali,

The quite different difficulties in the case of patulin ($C_7H_6O_4$) arose from a blindness to the multiplicity of possible hydrolytic ($H_3O^{\oplus}$)

pathways leading to the primary degradation products, [84] and HCOOH, which represent all the carbons of patulin and no change in overall oxidation state. The UV absorption at 276 mμ was observed but ignored in five of the six formulations of patulin which were variously proposed, all of which would account for these products as well as the reductive (H_2, HI, H_2) formation of β-methyl-γ-ethyl-butyrolactone. Patulin may thus be schematically represented as [85]. The fact that it is neutral and that acetylation shows it to have one hydroxyl group should lead to the correct formula [86], first proposed and later synthesized by Woodward. (∎)

3-15 Tetracyclines: Terramycin

Two classes of medically important natural antibiotics, the tetra-cyclines and macrolides, serve to represent the other end of the scale of molecular size and complexity in acetogenins. The first tetracycline structure to be unraveled, and later synthesized, was terramycin [87], the work of Woodward and the research group at the Chas. Pfizer Company, whose extensive degradations took place before the availability of NMR and mass spectra but were notable for their extensive use of other physical tools, especially IR and UV spectra, and pK measurements. Work centered around two series of degradation products, those from acid hydrolysis, all of which contained 1,8-dihydroxynaphthalene nuclei as shown by UV and the change in pK on boric acid complexation; and those from base, of which terracinoic acid and the simpler terranaphthol [88] were the key products. Strong acid produced terrinolide [89], the –$CONH_2$ group being identified by its loss in hot, strong H_2SO_4; this "decarboxamido" product was identified as a 1,2,4-trihydroxybenzene by comparison of its penta-methyl ether with UV models made by mixing a 1,8-dimethoxy-naphthalene with the various possible trimethoxybenzenes. The placement of the –$CONH_2$ between the resorcinol hydroxyls followed from the change in pK occasioned by its removal, again as compared to simple models.

Terracinoic acid ($C_{13}H_{12}O_6$) has three acidic functions (pK's: 2·6, 4·7, 9·5), methylation of only two of which (CH_2N_2) is reversible by saponification. It has also one ketone (cf. oxime formation) and one C—CH_3 and loses one molecule of CO_2 on pyrolysis to give a diacid (pK 4·7, 8·5). This suggests a salicylic acid because of the low pK lost on decarboxylation and the concomitant shift of a phenolic acidity

(9·5→8·5). The UV spectrum of the decarboxylated phenol is that of a 1-indanone with a conjugated –OH para- (not ortho)- to the ketone. The third pK (4·7) of terracinoic acid is that of a normal acid and its placement γ- to the ketone is shown by loss of the elements of HBr and CO_2 when mono-bromo-terracinoic acid is treated with carbonate.

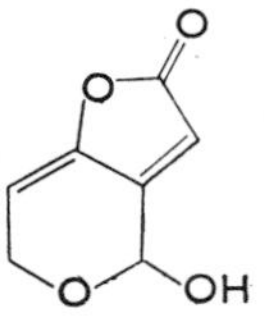

[86]
Patulin

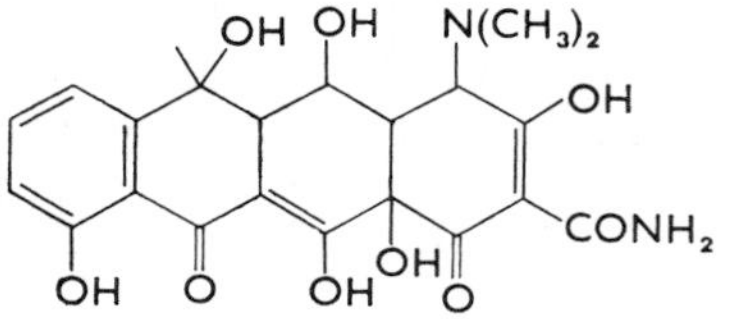

[87]
Terramycin

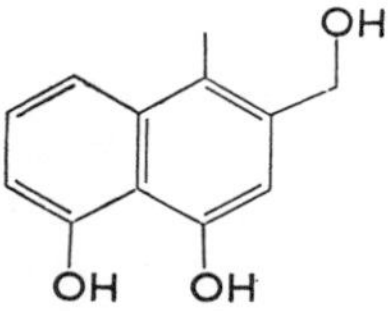

[88]

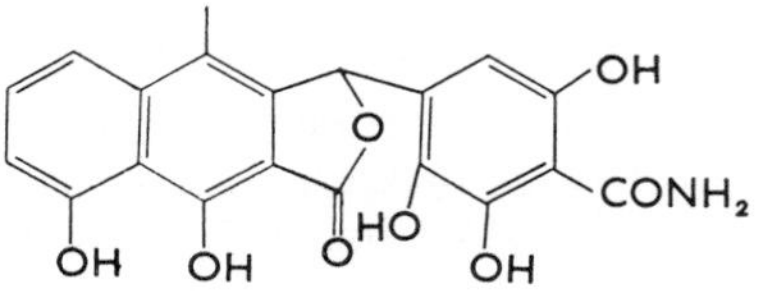

[89]

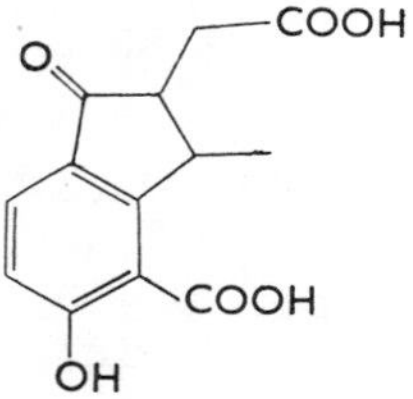

[90]

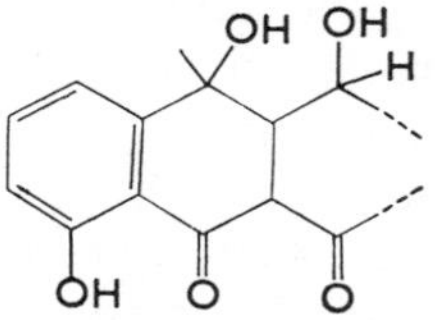

[91]

Since the IR spectra of terracinoic acid and its derivatives show the ketone in a five-membered ring and KOH fusion produces 2-hydroxy-6-ethyl-benzoic acid, terracinoic acid was accorded structure [90].

Terracinoic acid is racemic suggesting its asymmetric centers are not originally present in terramycin but rather are produced in the basic hydrolysis. An early partial structure [91] was therefore deduced for the left half of the terramycin molecule in order to account for the formation of both terracinoic acid and terranaphthol from the action of mild alkali on terramycin.

The deduction of the final structure of terramycin owed much to UV comparisons which established the 2-keto-1,8-dihydroxynaphthalene system in anhydroterramycin (ready dehydration of the benzylic tertiary hydroxyl in ring B) and the complex chromophore of the polyketone system on the underside by a number of adroit additions and subtractions of absorption curves from model o-hydroxyphenylketones. The same research group has more recently synthesized tetracycline and so confirmed structure [87].

3–16   Macrolides

The macrolide antibiotics, many of great commercial importance, are long chain molecules linked into a large-ring monocyclic lactone. The linear chains are usually liberally substituted with methyl groups, the biogenetic origin of which, in this unique class of natural substances, has been shown to be an admixture of propionate with acetate links in the forging of the long chain. A representative macrolide structure was deduced by Djerassi for methymycin [92], which is typical both in its functionality and in being a glycoside of an abnormal sugar, in this case desosamine. The sugar-free lactone, methynolide, afforded three compounds [93a], [93b], and [94], with $KMnO_4$, the largest (C_{16} out of C_{17}) of which [93a] was an α-hydroxy acid convertible to the second [93b] by $Pb(OAc)_4$. The ester [93b] was in turn saponified to 3-hydroxy-2-pentanone and an acid, the third $KMnO_4$ product [94], which, on pyrolysis to an α,β-unsaturated acid and ozonolysis, finally provided pyruvic and meso-α,α'-dimethylglutaric acids.

Since methynolide is an unsaturated ketone by UV, the mode of original $KMnO_4$ cleavage to [93a] with loss of one carbon and generation of two carboxyl functions is understandable (dots in [92]). The lactonization of one of these (δ-lactone by IR) also locates one of the hydroxyls in methynolide. Working backward now from the simplest to the most complex of these products allows us to deduce the structure of methynolide, which also provides the interesting mechanistic problem of the formation of [95] on KOH fusion of methymycin. (■)

The more complex macrolide, magnamycin [96], elucidated by Woodward, contains virtually every known functional group of oxygen. The epoxide of the doubly unsaturated ketone was shown by

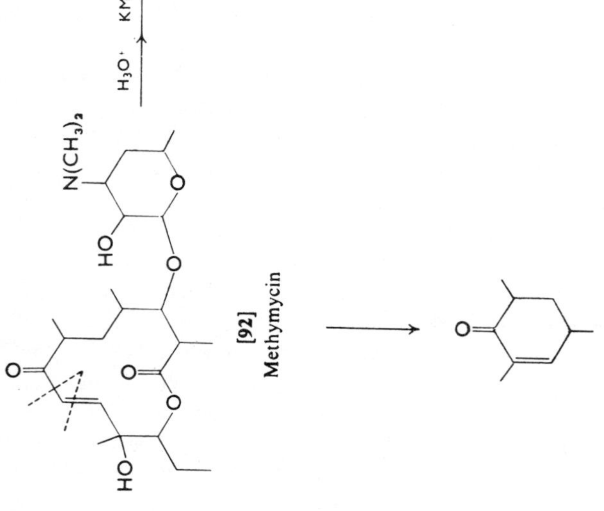

[94]

[93]

a, R = CH(OH)COOH

b, R = O

[92]
Methymycin

[95]

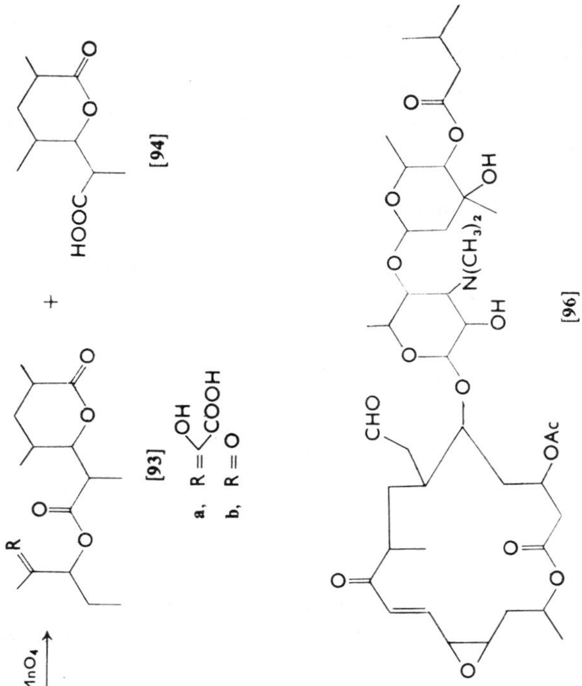

[96]
Magnamycin

reaction with KI/HOAC, which led simply to empirical loss of one oxygen atom and formation of the $\alpha,\beta,\gamma,\delta$-unsaturated ketone, shown by UV. Careful mild HNO_3 oxidations were instrumental in resolving the overall structure. Even the epoxide of maleic acid (from the magnamycin epoxy moiety) was isolated, and the obtention of pimelic acid from the tetrahydro derivative of the doubly unsaturated ketone was indicative of the unbranched chain down the left flank of the structure [96].

$$CH_3CH{=\!=}CH{-}C{\equiv}C{-}C{\equiv}C{-}CH{=\!=}CH{-}COOCH_3$$

[97]
Matricaria Esters

$$HC{\equiv}C{-}C{\equiv}C{-}CH{=\!=}C{=\!=}CH{-}CH{=\!=}CH{-}CH{=\!=}CH{-}CH_2COOH$$

[98]
Mycomycin

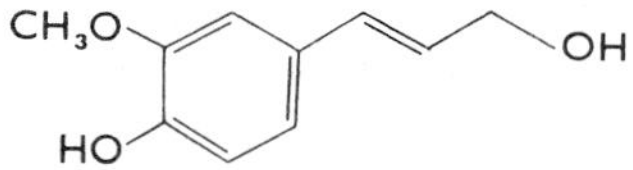

[99]
Coniferyl Alcohol

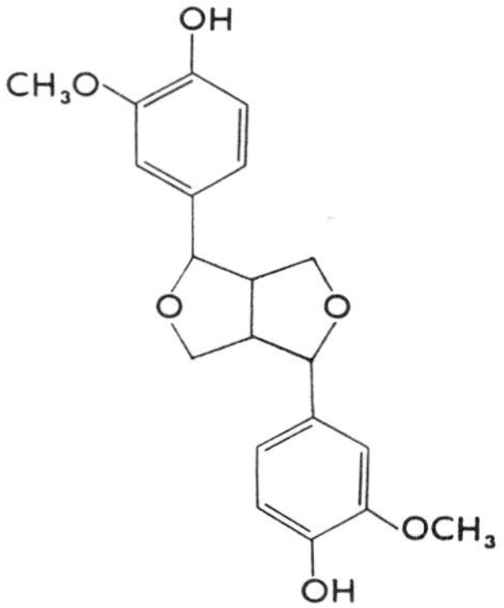

[100]
Pinoresinol

The basic linear chain of acetogenin biosynthesis is quite apparent not only in the macrolides but also in the simpler polyacetylenic natural products found in molds and in grasses, many of which have been elucidated by the Sörensons in Norway. The UV absorptions of these polyacetylene chromophores are highly characteristic and of enormous value in structural studies as may be seen in the example of the very common matricaria esters [97], hydrogenation of which requires six moles of H_2 and yields methyl decanoate. The UV spectrum places the unsaturation and the various possible *cis*- and *trans*-isomers, all found in nature, have all been synthesized. This group of natural products has generated much of the modern revitalization of acetylene chemistry. Mycomycin [98], an interesting fungal antibiotic, is optically active and the first natural allene known. Owing to the perpendicular arrangement (and consequent lack of overlap) of the two π-electron systems in the allene, the UV absorption of mycomycin is simply the sum of the chromophores

$$HC{\equiv}C-C{\equiv}C-CH{=}C \quad \text{and} \quad (CH{=}CH)_3.$$

The components of the musk glands (cf. muscone = 3-methylcyclopentadecanone) of certain animals, long prized as perfume constituents, are large-ring monocyclic ketones, presumably acetogenins, which, largely through the work of Ruzicka, laid the groundwork for our understanding of the chemistry of large-ring compounds.

3–17   Phenylpropanes: Podophyllotoxin

The phenylpropane compounds found in nature are either relatively simple molecules such as coniferyl alcohol [99], cinnamic acids and the dimeric lignans with ether rings like pinoresinol [100], or the exceedingly complex polymeric lignins, the fundamental structural constituents of wood. The phenylpropane dimers do, however, include a small group of somewhat more complex substances exemplified by the crystalline components of podophyllum resin. Podophyllotoxin ($C_{22}H_{22}O_8$) is a lactone, as evidenced by its saponification to a hydroxy acid (podophyllic acid), and contains three $-OCH_3$ groups (Zeisel) and one hydroxyl (from active hydrogen determinations and acetylation to a monoacetate). The hydroxyl is benzylic (not allylic; no double bond) since it is reversibly oxidizable to a conjugated ketone by MnO_2. These facts account for six of the eight oxygens; the other two and the full carbon skeleton are especially clearly deduced from oxidative cleavages.

Hot $KMnO_4$ yields 3,4,5-trimethoxy-benzoic and hydrastic [101] acids, while milder conditions afford [102] (R = COOH) and [103], the latter also being produced by reduction of the former while decarboxylation of the keto acid [102] (R = COOH) with Cu/quinoline leads to [102] (R = H), synthetically verified by a simple Friedel-Crafts reaction on methylenedioxybenzene. Taken together, these facts lead uniquely to a partial structure [104]. Mild dehydrogenation

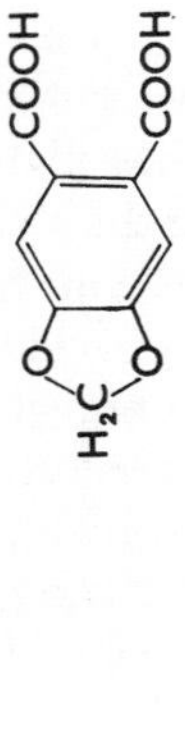

[101]

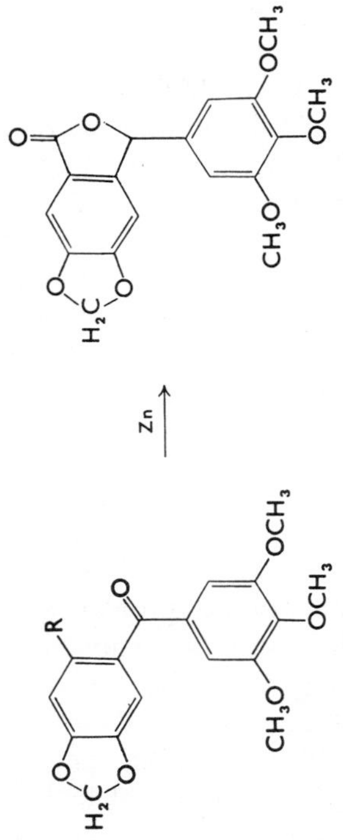

[102]

[103]

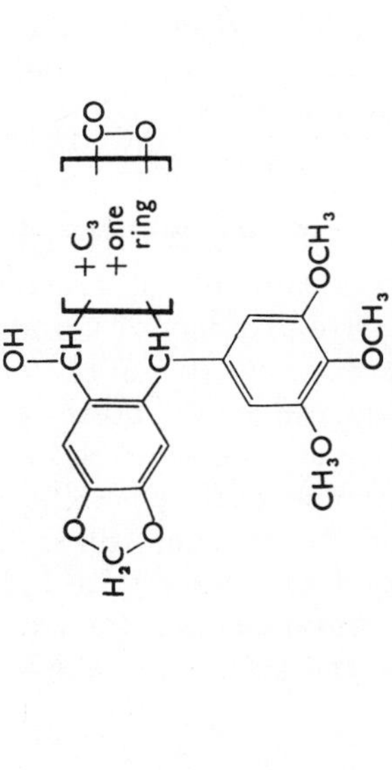

[104]

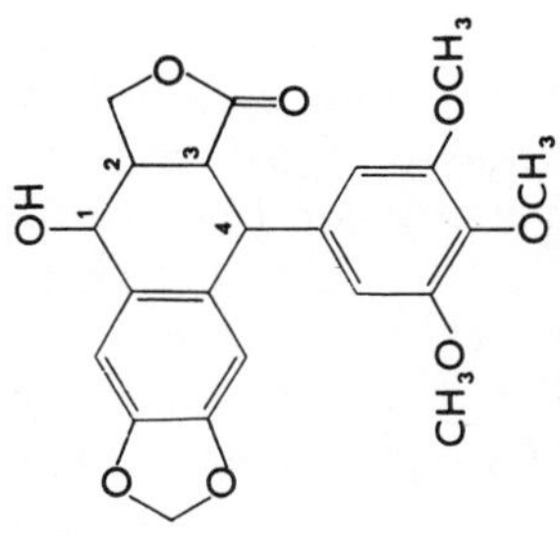

[105]
Podophyllotoxin

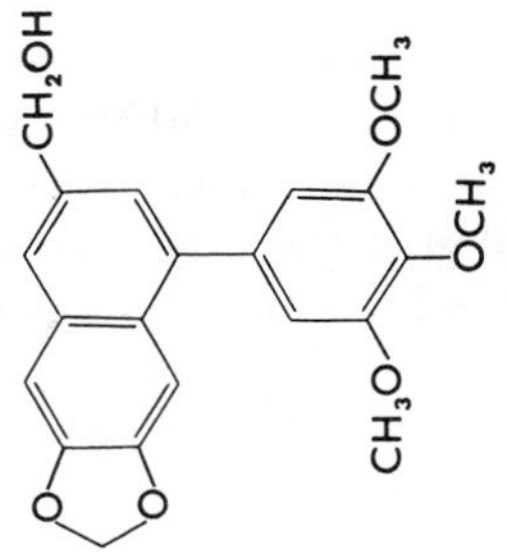

[106]

of podophyllotoxin affords the lactone $C_{22}H_{18}O_7$ (loss of $H_2 + H_2O$) which is transformed by $KMnO_4$ into benzene pentacarboxylic acid. The formula for podophyllotoxin must therefore be either [105] or an isomer with the lactone orientation reversed. Confirmation of [105] comes from oxidative decarboxylation of podophyllic acid to [106] and its destructive oxidation to benzene-1,2,3,5-tetracarboxylic acid.

The stereochemical problem is an especially interesting one here because of the four adjacent asymmetric centers which imply eight possible diastereomers. Podophyllotoxin acetate is much more readily converted into the 1,2-olefin by pyrolysis than its epimer at C-1 (made by two successive displacements—PBr_3, then H_2O/Na_2CO_3—similar to Walden's famous original study of configurational inversion and retention in the natural malic acids). Since acetate pyrolysis is a *cis*-elimination, the substituent groups at C-1 and C-2 are *trans* in podophyllotoxin. The $\Delta^{1,2}$-olefin thus formed may be isomerized to the $\Delta^{2,3}$ olefin by mild base and on to the $\Delta^{3,4}$-isomer by strong alkali. Hydrogenolysis of podophyllotoxin yields the C-1 desoxy derivative.

We may now construct the following stereochemical problem and solve it for the configuration of podophyllotoxin if we assume *cis*-introduction of hydrogen in catalytic hydrogenation and epimerization in mild base only adjacent to the lactone carbonyl. Label the four possible diastereomers of the C-1 desoxy-lactone A, B, C, and D. Let A be the one formed directly from the natural product by hydrogenolysis of the secondary, benzylic alcohol. Catalytic hydrogenation of the $\Delta^{1,2}$-olefin yields both B and C while hydrogenation of both the $\Delta^{2,3}$- and $\Delta^{3,4}$-olefins yields only D. Treatment with mild base converts A to C and D to B.

3–18   Acetogenin Alkaloids; Annotinine

The curious family of structures elaborated by the Lycopodium mosses are unlike any of the alkaloids produced by higher plants. After the extensive studies of Wiesner's group establishing the structure of annotinine, the other structurally related alkaloids followed much more easily by analogy. The annotinine elucidation provides a happy example of the possibility, in appropriate cases, of unraveling much of the skeleton through functional group interrelations, especially with the ability to determine carbonyl ring size from IR spectra. Annotinine ($C_{16}H_{21}NO_3$) contains a tertiary nitrogen, a C—CH_3, and an epoxide which is reversibly interconvertible with a chlorohydrin. Oxidation of annotinine with mild $KMnO_4$ converts the basic nitrogen to a lactam, the carbonyl of which is seen to be adjacent to the epoxide as shown by the ready dehydration of the chlorohydrin

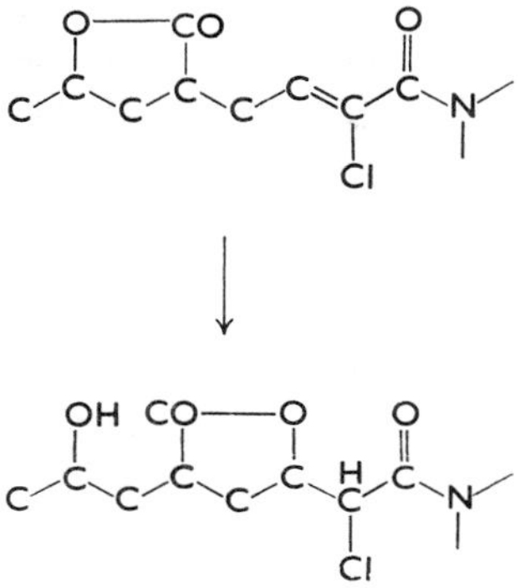

[107]

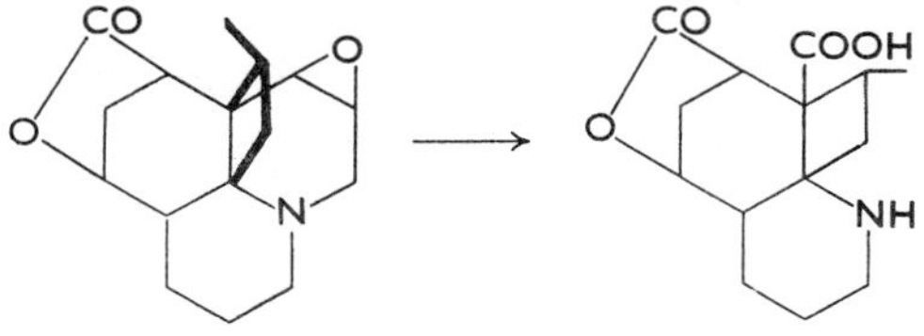

[108]
Annotinine

[109]

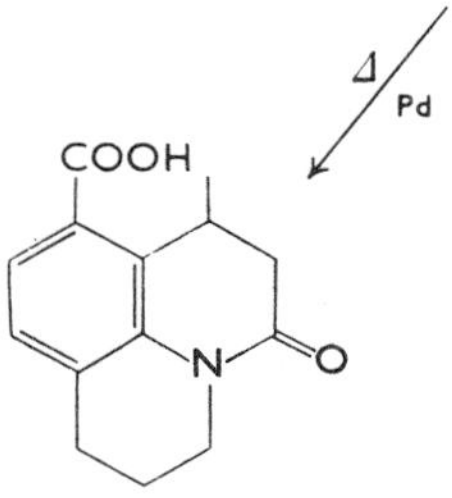

[110]

to an α-chloro-unsaturated lactam (C—CH=CCl—CO—N–), identified by its UV and IR absorption. Annotinine also contains a γ-lactone (by IR) which, in the case of the unsaturated lactam derivative, can be opened to a hydroxy acid and reclosed to an isomeric

γ-lactone with no UV absorption. This interrelation allows the partial structures [107] for the two lactones and incidentally develops much of the carbon skeleton. (■) The full structure [108] for annotinine was deduced from further chemical evidence and confirmed by X-ray analysis. $KMnO_4$ oxidation of the first unsaturated lactam [107] yields [109], which in turn undergoes some marvellous mechanistic pyrotechnics on palladium dehydrogenation to [110]!

4

■ TERPENES AND STEROIDS

FROM ANTIQUITY the essential oils of many plants have been prized for their varied and fragrant odors. They were originally separated by gentle heating and collected as crude distillates for use, then as now, in perfumery. Later the less destructive steam distillation procedure was devised as an improvement and by the late Renaissance as many as sixty essential oils were named and in use. As these distillates were usually, of course, complex mixtures, early terpene studies were plagued by the variance in homogeneity (not always appreciated by the investigators), by inadequate fractional distillation equipment, and by the rarity of crystallizable derivatives. In contemporary practice with volatile terpenes the homogeneity problem is usually solved by gas chromatography. Historically, the first major advance was the introduction by Tilden (1877) of the reaction of NOCl to produce crystalline derivatives of olefinic terpenes (see Section 4–2, limonene).

Early terpene work, notably that of Wallach, in the decades around the turn of the century was mostly done on monoterpenes partly because of their volatility, which allowed them to be separated easily from the many more complex products of the plant, and partly because of their relative molecular simplicity. Meaningful studies in the higher terpenes mostly date from about 1920 and are dominated by the brilliance and diligence of Ruzicka and his collaborators in Zürich. Although the steroids were increasingly suspected of biogenetic kinship with the terpenes, they remained the subject of a separate line of endeavor through the major classical period of structure determination and were not demonstrated to have a terpenoid origin until the 1950's.

4–1   Important Elucidation Techniques

Two invaluable tools form the foundation for elucidating the skeletons of the terpene molecules: the isoprene rule, outlined in Chapter 2, and the use of sulfur dehydrogenation. This reaction was used by Vesterberg in 1903, but its broad applicability was first recognized and systematically exploited in the early 1920's by Ruzicka, who applied it to the lower terpenes (C_{10}, C_{15}, C_{20}), synthesizing the aromatic products to establish their identity. In essence the method involves heating the unknown compound with sulfur (or selenium) and distilling out the products—aromatic compounds representing a dehydrogenation of the original molecular skeleton to its nearest fully aromatic relative(s). This is illustrated in Chart 1 which shows the major dehydrogenation products of the main families of terpene skeletons (cf. Chart 6, p. 34).

It was, in fact, the recognition of a common dehydrogenation product which first allowed the grouping of many terpenes into families bearing common carbon skeletons and so enormously aided their interpretation by correlation with each other. As the chart implies, quaternary methyls are usually lost in dehydrogenation but sometimes migrate to an adjacent site and are then retained. An elegant use of the reaction involves locating a functional group by "marking" its site with a methyl group (often by the reaction of a ketone with CH_3MgBr) before dehydrogenation and then demonstrating the position of the new methyl on the derived aromatic skeleton by appropriate degradations and synthesis.

The value of the method was surprisingly not early recognized in the steroid field. Diels first used it on cholesterol in 1927, obtaining chrysene (1,2-benzphenanthrene) and the "Diels hydrocarbon"

(Chart 1). The chrysene clue was central to the revisions of 1932 leading to the correct formula (see p. 130) but Diels' even more significant cyclopentano-phenanthrene remained itself unsolved until its synthesis in 1933. It then became the talisman for the recognition of the steroid skeleton in scores of new natural products.

4-2   Monocyclic Monoterpenes: Limonene and Carvone

The characteristic odor of lemons is largely that of the major volatile component, limonene [1], one of a number of monocyclic monoterpene dienes which are readily transformed into *p*-cymene on dehydrogenation. A crystalline derivative of the oily limonene arises from reaction

CHART I •

Major Terpene Dehydrogenation Products

Monoterpenes:

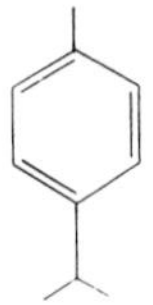

p-Cymene

Sesquiterpenes:

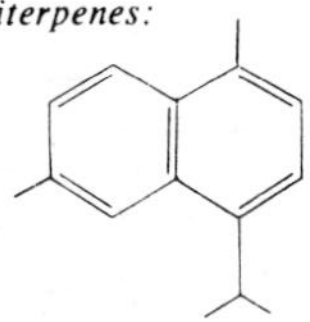

Cadalene

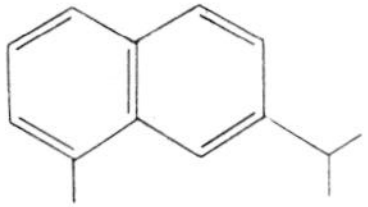

Eudalene

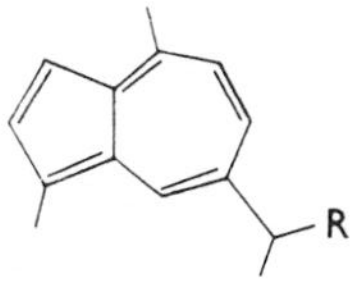

S-Guaiazulene (R = CH₃)
Chamazulene (R = H)

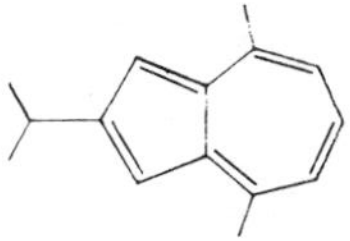

Vetivazulene

Diterpenes:

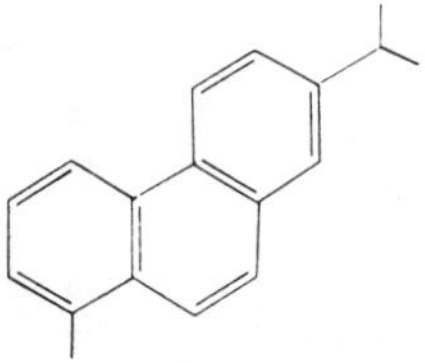

Retene

and

as well as

HO

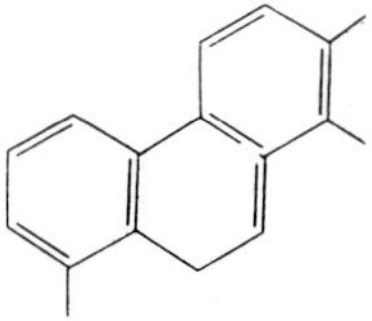

Steroids

"Diels Hydrocarbon"

Pentacyclic Triterpenes:

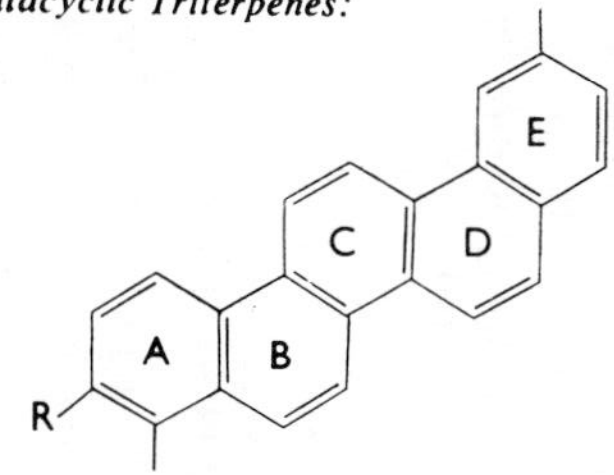

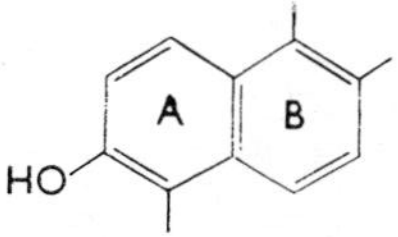

Dimethylpicenes
(R = H or OH)

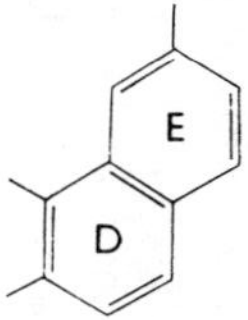

Tetracyclic Triterpenes:

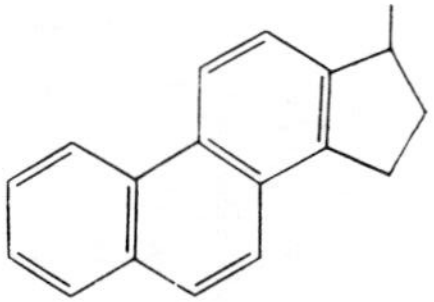

with NOCl; the simple addition product formed initially [2] tautomerizes to the more stable chloro-oxime [3]. Mild treatment with base then yields carvone oxime which, in turn, may be hydrolyzed to carvone [4], the major constituent of caraway and dill oils. The correlation thus established between these two monoterpenes is confirmed by Wolff-Kishner reduction of carvone to limonene. The location of the second double bond in each then follows from sodium amalgam reduction to convert the unsaturated ketone to saturated alcohol and oxidation of the remaining double bond, which yields a methyl ketone with one less carbon [5]. The position of the keto-group was ascertained by an acid-catalyzed shift of the isopropenyl double bond into the ring to form carvacrol [6] which was identified by synthesis.

An interesting side problem developed in the carvone work which serves to exemplify the common occurrence of such happenstance and its value in enriching the mainstream of organic chemical understanding. Carvone readily adds HBr to the terminal double bond and is thus converted to a tertiary bromide, on which the action of base, eliminating HBr, affords eucarvone, a ketonic isomer of carvone first prepared by Baeyer. For many years a controversy over its formulation as [8] (Baeyer) or [9] (Wallach) was fed by conflicting chemical evidence such as ozonolysis to caronic acid (*cis*-3,3-dimethylcyclopropane-1,2-dicarboxylic acid) and $KMnO_4$ oxidation to α,α-dimethylsuccinic acid. The benzylidene derivative of eucarvone, cited by Wallach in favor of [9], is actually [10] and eucarvone exchanges *three* hydrogens for deuterium when equilibrated with $C_2H_5O^{\ominus}/C_2H_5OD$. In cases of this kind involving tautomerism during chemical reactions, only physical evidence on the resting molecule can lead to an unequivocal formulation. One of the first important applications of NMR spectroscopy to structure problems was Corey's demonstration that eucarvone is actually [9], with three olefinic hydrogens.

4–3 Bicyclic Monoterpenes: Pinene and Camphor

The bicyclic monoterpenes, distinguished by their many, often subtle, skeletal rearrangements, provided the first, and for a long time the major, framework for the developing study of these intriguing reactions. The extensive studies currently being pursued in the area of nonclassical carbonium ions still utilize the strained ring systems of those monoterpenes, notably the norbornane skeleton which takes its

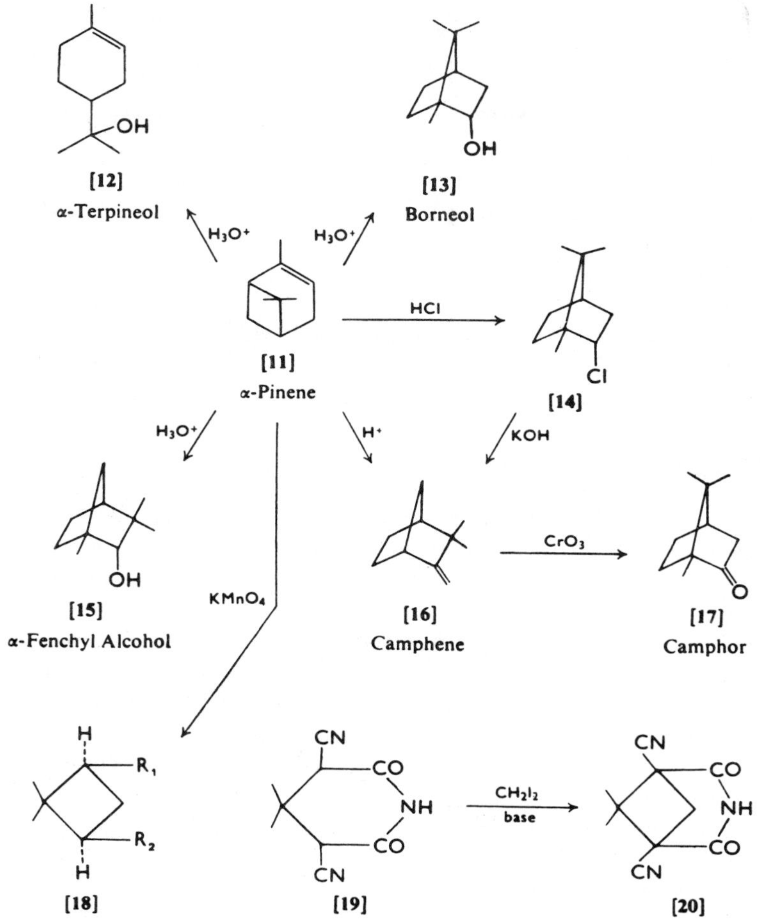
OH
[12]
α-Terpineol
OH
[13]
Borneol
H₃O⁺
H₃O⁺
HCl
[11]
α-Pinene
Cl
[14]
H₃O⁺
H⁺
KOH
OH
[15]
α-Fenchyl Alcohol
KMnO₄
CrO₃
[16]
Camphene
O
[17]
Camphor
H
R₁
R₂
H
CN
CO
NH
CO
CN
CH₂I₂
base
CN
CO
NH
CO
CN
[18]
[19]
[20]

a: $R_1 = COCH_3$; $R_2 = CH_2COOH$
b: $R_1 = COOH$; $R_2 = CH_2COOH$
c: $R_1 = COOH$; $R_2 = COOH$
d: $R_1, R_2 =$ —CO—O—CO—

name from borneol [**13**] without its methyl substituents. That these rearrangements created many obstacles to simple structure proof in the beginning is dramatized by the many different acid-catalyzed reactions of α-pinene, grouped around formula [**11**]. The products are all rearranged and are all themselves natural terpenes!

Pinene, the most important and widespread terpene hydrocarbon is the major constituent of turpentine (from which the word terpene was derived), the volatile component of pine resins valued for centuries as a solvent. Pinene was first analyzed by Lavoisier and its correct formula was among the first results of Dumas' important molecular weight method. Biot's first observation of optical activity (1815) was made on turpentine. Skeletal rearrangement was first recognized in pinene by Berthelot (1862), and it was Wagner who first enunciated the correct structure.

The structure of α-pinene [11] (β-pinene is the isomer with the double bond exocyclic) was convincingly demonstrated by Baeyer with a $KMnO_4$ oxidation to a methyl-ketone acid [18a] which was oxidized by hypobromite to bromoform and pinic acid [18b], and this in turn oxidized further to a *cis*-norpinic acid [18c], a diacid containing only one less carbon. A *trans*-norpinic acid was later synthesized neatly by the addition of methylene iodide to the sodio-derivative of [19] and acidic hydrolysis-decarboxylation of the product [20]. The *trans*-acid is convertible on heating to the *cis*-anhydride [18d] which in turn yields the isomeric *cis*-acid on mild hydrolysis. The starting material [19] may be made in one step from acetone, cyanoacetic ester and ammonia.

Long prized for its presumed medicinal properties and now commercially important in plastics and photography, the monoterpene camphor was a major subject of terpene chemistry in the nineteenth century. About thirty different constitutions were proposed for this C_{10}-ketone before Bredt's correct formulation in 1893! Only rarely appearing in natural sources, camphor is synthesized commercially from α-pinene in three steps, each one of which involves a skeletal rearrangement: HCl yields bornyl chloride [14], which is transformed into camphene [16] by KOH and this in turn into camphor [17] by CrO_3, which initiates a Wagner-Meerwein rearrangement to iso-borneol prior to oxidation.

4–4   Sesquiterpenes

Until Ruzicka's work in the early 1920's the correct formula of only one sesquiterpene (farnesol) was known despite extensive studies of these compounds in previous decades. Most sesquiterpenes are olefinic hydrocarbons (or the alcohols derived from them by simple hydration), and their structures usually follow from dehydrogenation, the isoprene rule, and subsequent oxidations which open ethylenic linkages to determine their locations on the skeleton. As many of these compounds can be interrelated chemically, the structure problems are often much simplified. Thus many of the simple monocyclic sesquiterpenes and those with cadinane and eudesmane skeletons (see Chapter 2—Chart 6) were elucidated by 1940 while structures for the perhydroazulene sesquiterpenes have been more recent and usually attended by more problems. A central feature of the structure proof of guaiol [21] was the clever conversion to a hydronaphthalene via ozonolysis of [21] to the expected diketone followed by internal aldol cyclization of this to [22], which then yielded cadalene (Chart 1) on dehydrogenation.

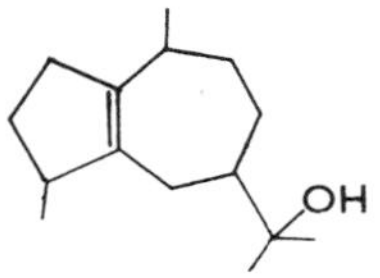

[21]
Guaiol

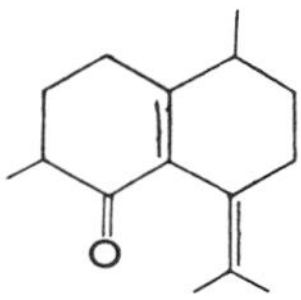

[22]

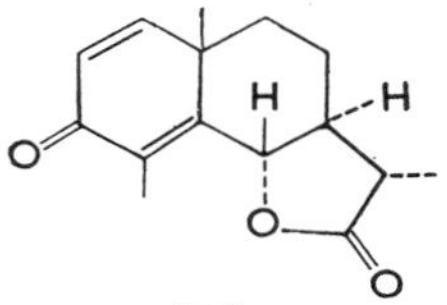

[23]
Santonin

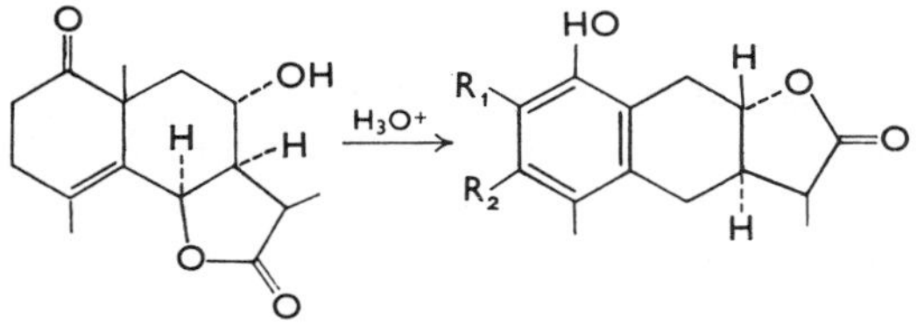

[24]
ψ-Santonin

H_3O^+

[25]
a: $R_1 = CH_3$; $R_2 = H$
b: $R_1 = H$; $R_2 = CH_3$

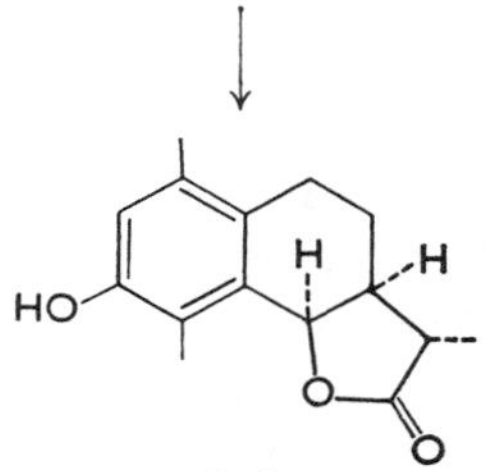

[26]
(−)-α-Desmotroposantonin

A. SANTONIN   The various species of *Artemisia* provide the unique quality of absinthe as well as a rich harvest of sesquiterpenes, the most extensively investigated being santonin [23]. Santonin was first isolated in 1830 and has long been widely used in India to combat intestinal worms. In the nineteenth century Italian investigations of santonin led to its recognition as an unsaturated keto-lactone. Its easy conversion to aromatic compounds such as desmotroposantonin [26] led to the belief that the ketone and olefinic groups were in the same ring; hence early formulations utilized the skeleton of desmotroposantonin, placing a dienone in the dimethylated ring. Such formulas, however, were simply ketonic tautomers of phenols and thus unacceptable now and at least suspicious then. The correct formula was unveiled by Clemo in 1929 when he observed that none of the existent formulas fitted the isoprene rule and that the properly terpenoid eudesmane skeleton also afforded, in its quaternary center, an explanation for the existence of a dienone in one ring that could not tautomerize to a phenol. The methyl migration required to furnish desmotroposantonin, and other products of the same skeleton, was the first example of the now familiar dienone-phenol rearrangement and desmotroposantonin was synthesized by Clemo in partial verification of his views.

[27] [28] [29]
Longifolene

In cold aqueous H_2SO_4 santonin yields $(-)$-α-desmotroposantonin via methyl rearrangement and subsequent epimerization of the benzylic center to form the more stable *cis*-fusion of the γ-lactone ring. In stronger acid this first product is converted to the isomer $(+)$-β-desmotroposantonin, which then yields $(+)$-α-desmotroposantonin in hot alkali. Thus in two reactions $(-)$-α-desmotroposantonin has been transformed into its $(+)$-mirror image and the same sequence of strong acid, then alkali, of course reconverts the $(+)$-back to the original $(-)$-enantiomer, completing a cycle. In this rare and intriguing cycle, then, all three centers of asymmetry must be inverted. The two involved in the *cis*-lactone ring fusion are probably removed and reformed (still *cis* but inverted) by way of acid-catalyzed elimination to the styrene, while alkali, in its turn, epimerizes (via the enolate) the methyl adjacent to the lactone carbonyl. Dauben showed that the similar sesquiterpene ψ-santonin [24] affords the two phenols [25] by analogous but even more splendid acid-catalyzed rearrangements.

This series was not the only one to confuse the early workers, for, while acid reagents almost invariably cause rearrangement to an aromatic substance, vigorous treatment of santonin with alkali yields santonic acid with a composition ($C_{15}H_{20}O_4$) empirically representing the addition of a mole of water as expected for a lactone. But santonic acid is a *saturated* acid with *two* ketone groups and therefore tricyclic! (■) The solution of the problem was offered by Woodward 75 years after its first isolation by Cannizzaro in 1873. Woodward argued that, after lactone cleavage, the tetrasubstituted double bond was isomerized by alkali to the β,γ-position, thus creating an enol which tautomerized to a 1,4-diketone and subsequently underwent an internal Michael addition [1] onto the remaining double bond. Santonic acid was therefore formulated as [30], suitably explaining all the degradative chemistry which had collected on this curious acid. A further amusing transformation product was also elucidated at about the same time: santonic acid is converted into santonide [31], an isomer of santonin, by heating in acetic acid, then pyrolyzing. The mechanism of this change is also most unusual and a challenge to the student of nonclassical carbonium ions!

B. PYRETHROSIN   Pyrethrosin, the bitter constituent of pyrethrum flowers was shown to be $C_{17}H_{22}O_5$, composed of a γ-lactone, an acetoxy group, and two double bonds. The fifth oxygen was originally believed to be a secondary alcohol since CrO_3 afforded a ketone $(C_{17}H_{20}O_5)$, and, although acetic anhydride in pyridine had no effect on pyrethrosin, acid-catalyzed acetylation gave an acetyl derivative. Finally, since an azulene had been formed on dehydrogenation, pyrethrosin was awarded a bicyclic formula of the guaiazulene skeleton. However, Barton found that the compound had no active hydrogen by D_2O exchange and hence could not contain the postulated hydroxyl. Accordingly, there must be a reactive ether present and therefore a *monocyclic* carbon skeleton. Furthermore, both the CrO_3 and acid-catalyzed acetylation products were shown to have bicyclic carbon skeletons, and these of the eudesmane rather than guaiane kind, by conversion to the compound [33] which was readily interrelated with ψ-santonin [24]!

On the basis of this and other information (e.g., conversion of the epoxide to a ketone by BF_3 in the tetrahydro compounds), pyrethrosin was formulated as the first ten-membered monocyclic sesquiterpene lactone [32]. Its cyclization by acid media to cyclopyrethrosin (arrows) with the eudesmane skeleton, or by dehydrogenation to an azulenic skeleton, reflects the postulated biogenesis of these sesquiterpenes (Chapter 2). The same dual course of cyclization had thwarted

---

[1] Corey's economical synthesis of the sesquiterpene longifolene [29] utilized the conception of this internal Michael addition in the key reaction: [27] → [28].

earlier attempts to reach an acceptable structure for the terpene germacrone [34], which yields guaiazulene on dehydrogenation but eudesmane on catalytic hydrogenation in acid.

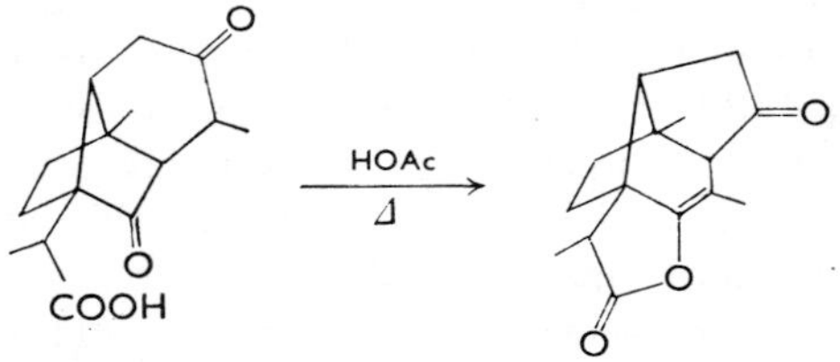

[30]
Santonic Acid

[31]
Santonide

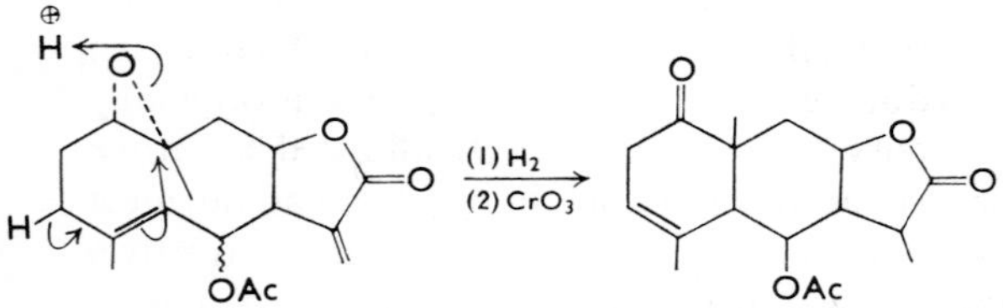

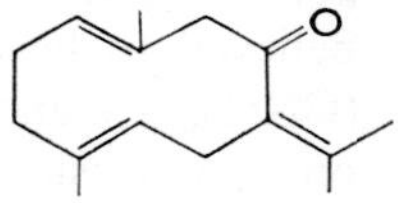

[32]
Pyrethrosin

[33]

[34]
Germacrone

C. β-VETIVONE   The case of β-vetivone ($C_{15}H_{22}O$), the ketonic main
constituent of oil of vetiver, affords an especially elegant and simple
structure proof. On catalytic hydrogenation it forms a hexahydro
compound (a saturated secondary alcohol oxidizable to a ketone which
forms a dibenzylidene derivative) as well as a tetrahydro derivative,
also formed by sodium amalgam reduction, which on ozonolysis yields
acetone and a hydroxy-ketone, $C_{12}H_{20}O_2$, with two C-methyl groups
and its ketone in a five-membered ring (by IR). While vetivone is
optically active, the tetrahydro compound is inactive. This salient
observation reduces the structure problem to only three possibilities!
(■)

Taking as a starting point the plane of symmetry that must be present
in the (*meso*)tetrahydro compound, we may construct on it all the
known groupings in such a way that all atoms on one side of the mirror
plane are matched by the same on the other, and with the result a

bicyclic skeleton. Such a construction must result in [35], and the structure of β-vetivone [36]—except for the relative configuration of the methyl—is established by its ultraviolet spectrum (λ_{max} 238 mμ) indicating a disubstituted α,β-unsaturated ketone. (In Plattner's original elucidation, he also knew fairly early that vetivazulene was formed on dehydrogenation, although on the other hand he did not possess the convenient spectral interpretations.)

D. TENULIN The most widespread perhydroazulene sesquiterpenes, however, yield guaiazulene, first obtained from guaiol (or chamazulene in those cases in which the familiar isopropyl pendant exists as a lactone as in santonin [23]). The lactone matricin [37] which is in the same oxidation state as an azulene actually turns blue with the formation of guaiazulenic acid [38] merely on warming briefly in aqueous acid.

The constituents of the sneezeweeds (*Helenium* species) include a number of bitter substances which yield chamazulene on dehydrogenation and structures for tenulin and helenalin were first proposed with the guaiazulene skeleton since it also fits the isoprene rule. However, in an extensive investigation of these compounds, which has interrelated them all and produced over a hundred characterized derivative compounds, Herz discovered that in appropriate derivatives, the NMR signals from the five-ring methyl group were unsplit by adjacent protons so that the methyl must instead occupy a quaternary position. His correct formula for tenulin [39] not only demonstrates the diabolism of nature in providing an exception to the isoprene rule which *also* rearranges on dehydrogenation, but also exhibits a remarkable functional group in the cyclic hemiacetal which exists so tenuously that boiling in tap water serves to open the ring to an acetoxy isomer (isotenulin) by the retroaldol mechanism summarized in the arrows shown on [39]. When Barton first formulated this hemiacetal grouping he neatly showed that the retroaldol reaction was activated by the lactone carbonyl and not the unsaturated ketone by the demonstration that the reaction also occurred on dihydrotenulin and its oxime.

Among the many interesting reactions of tenulin is its conversion to desacetyl-neotenulin [40] on treatment with aqueous base, the mechanism of which may be left to the reader's fancy. The enormous skein of reactions that have been carried out on the *Helenium* sesquiterpenes may well now provide a platform for studies in the conformational

analysis of cycloheptanes as the steroids did for cyclohexanes. It may
be noted that tenulin, like cholesterol, has eight asymmetric centers.

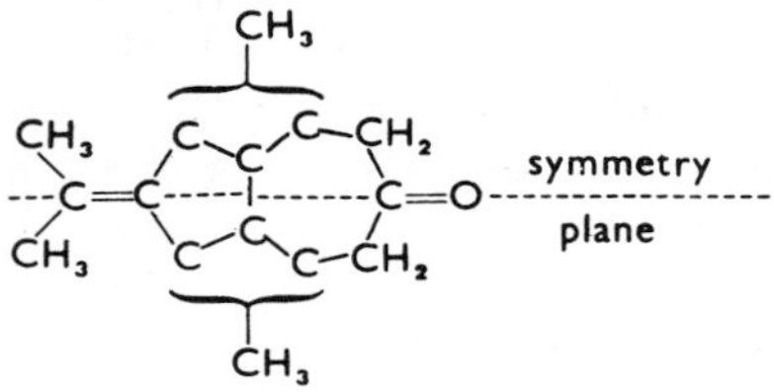

[35]

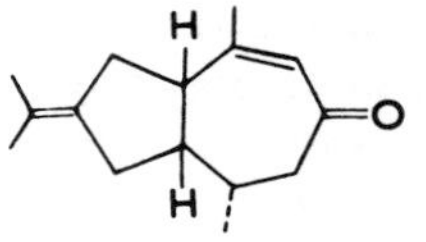

[36]
β-Vetivone

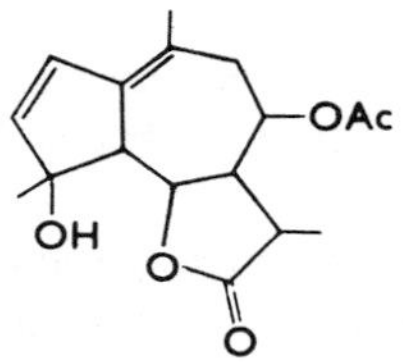

[37]
Matricin

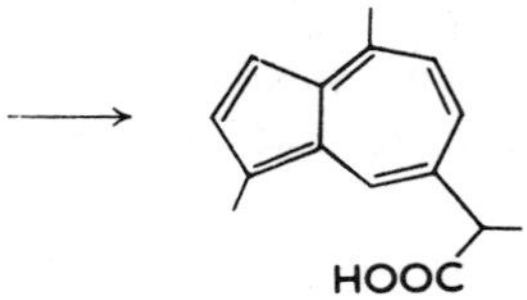

[38]

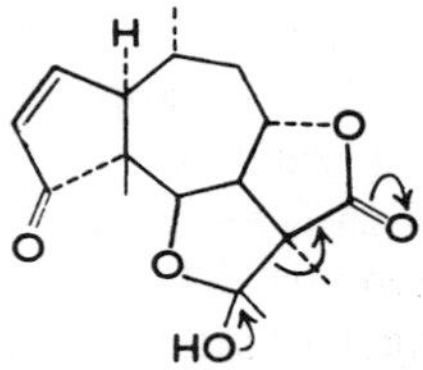

[39]
Tenulin

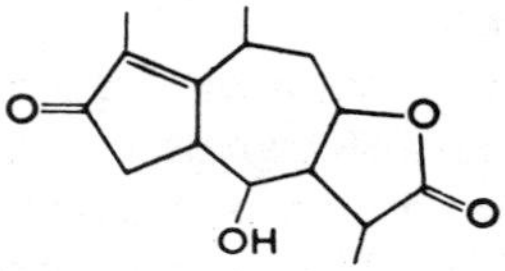

[40]

E. CARYOPHYLLENE   Several more complex sesquiterpenes do not
belong to the structural families of the foregoing examples, and their
elucidation was far more burdensome since they did not yield any
meaningful dehydrogenation products. This may readily be seen in
the structures for cedrol [41] from cedarwood oil and longifolene [29].
The former was the subject of extended progressive degradations by
oxidation leading finally to its structure and synthesis by Stork, while
the structure of longifolene was finally deduced only by X-ray crystallographic analysis despite much chemical work which had afforded no
solution.

   Among the many natural constituents in oil of cloves is the bicyclic
diene, caryophyllene, $C_{15}H_{24}$. That one of the rings was a cyclobutane

was shown by oxidation to a series ([42]; $n = 0, 1, 2$) of diacids which were degraded further and ultimately confirmed by synthesis. The environs of the two (unconjugated) double bonds were deduced to be as in [43] by ozonolysis to the diketo-acid [44] which underwent a facile aldol condensation to [45], shown to be a methyl ketone by NaOBr oxidation to bromoform and the corresponding unsaturated C_{13} diacid. Since caryophyllene has no other C—CH_3 groups (■) the structure [46] may be uniquely derived (except for the *trans* nature of both the double bond and the ring fusion, deduced separately).

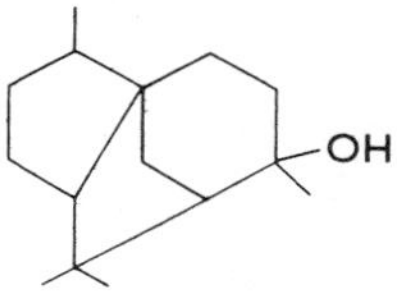

[41]
Cedrol

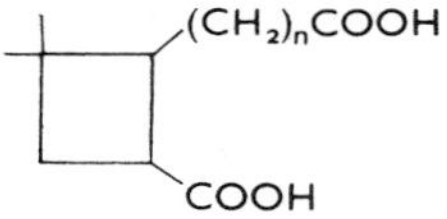

[42]

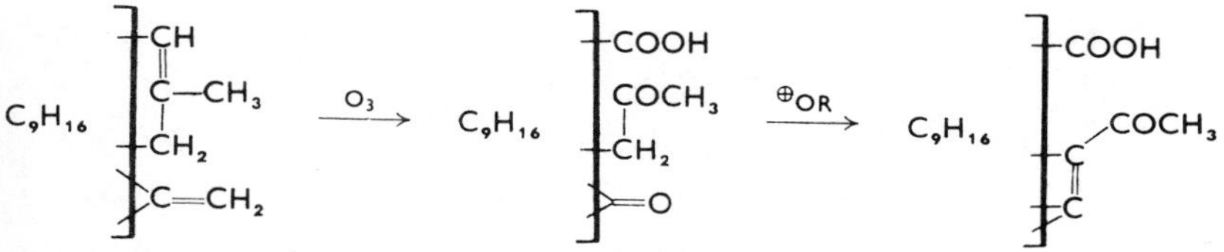

[43]   [44]   [45]

Separate structural problems were concurrently provided by the products of acid-catalyzed cyclization of caryophyllene, viz., clovene [47] and β-caryophyllene alcohol [48]. In an elegant synthesis of caryophyllene, Corey executed a reversal of this common cyclization of the nine-membered ring to tricyclic derivatives, cleaving the synthetic [49] with base, as shown, to a ketone, the conversion of which to =CH_2 produces caryophyllene.

4–5 Diterpenes

The most common of the smaller group of diterpenes is abietic acid [50], the chief constituent of pine rosin. Abietic acid is not a true terpene but is produced by a methyl migration from natural acids with the skeleton of dextropimaric acid [2–73]. These and other diterpene "resin acids" have been studied for over a century owing to their commercial interest and ready availability. Retene was produced from them by sulfur dehydrogenation (Vesterberg, 1903) in practically the first use of that tool but its elucidation was long and difficult as it required developing *de novo* much of the chemistry of phenanthrenes. This diterpene degradation product has also been isolated from fossilized pine trunks!

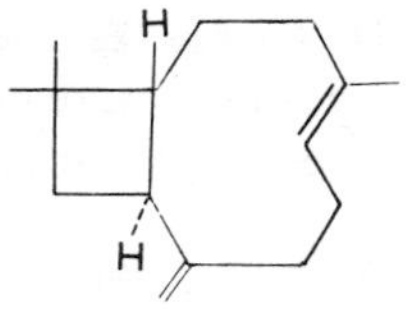

[46]
Caryophyllene

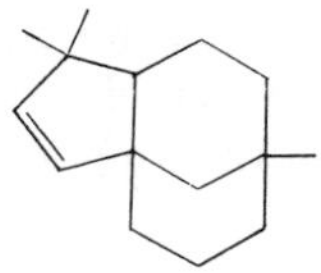

[47]

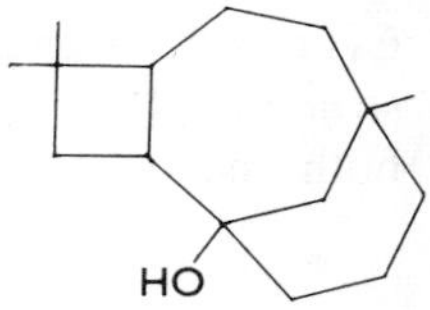

[48]

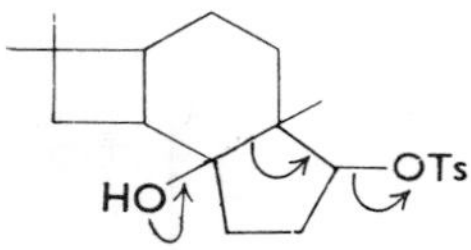

[49]

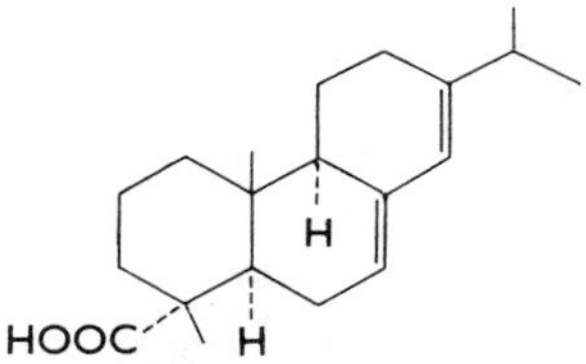

[50]
Abietic Acid

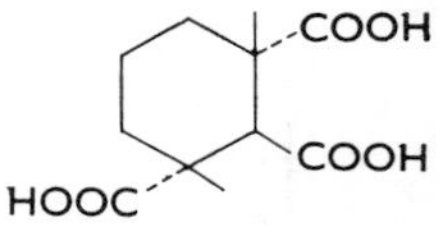

[51]

Retene contains eighteen of the twenty carbons, however, and one other was known to be lost as CO_2; the remaining one was assumed by Ruzicka to be an angular methyl on the basis of his experience with dehydrogenation. He then obtained from ring A of abietic acid by vigorous $KMnO_4$ oxidation the triacid [51], which contains both these two missing carbons. Barton later demonstrated the stereochemistry illustrated for [51] by a study of its dissociation constants (pK) vis-à-vis model acids and the fact that it is optically inactive and so has a mirror plane. This immediately accounts for the configurations of three of the four centers in abietic acid.

4–6 Diterpene Alkaloids

More complex diterpenes are to be found in the potent plant-growth stimulator, gibberellic acid [52], and a number of natural furans such as columbin [53], the latter exhibiting a skeleton with rearranged methyl groups such as are more often found in triterpenes. The other major and complex group of diterpenes is the collection of diterpene alkaloids generally found in delphinium and larkspur, which had a vogue as murder poisons in the last century; pseudoaconitine is lethal to mice in amounts as small as 0·025 mg/kg of body weight. Much early work in the last century and by Jacobs in the 1930's led to no structures owing to the reluctance of these terpenes to produce any dehydrogenation products of deductive value. Although Jacobs believed his alkaloids to be diterpenes, it was only with the later discovery and elucidation of the structurally simpler isomers garryine and veatchine [54] by Wiesner that clarification began to emerge.

It was early recognized that both of these alkaloids contained an oxazolidine ring, attached at either side of the nitrogen as in [54] and the alternate (→) in garryine, since they were interconvertible (acid

[52]
Gibberellic Acid

[53]
Columbin

opens the ring to a β-hydroxylethylimmonium salt) and formed a common dihydro product (a β-hydroxyethylamine). The two oxazolidine carbons and oxygen are removable in various pyrolytic reactions which yield the six-membered imines (the ring size is known from the IR absorption of the corresponding lactam). The other functionality consisted of the allylic alcohol (see [54]), isomerizable to an α-methyl-cyclopentanone (ring size by IR again).

In these alkaloids selenium dehydrogenation was successful and led to the two aromatics [55] and [56], both identified first by their characteristic UV chromophores, then by comparison with authentic synthetic samples. Many of the atoms in these products must be common to both and in fact lead only to the skeleton in [54] for the alkaloid, the quaternary atoms occurring at the sites where cleavage must occur on dehydrogenation to the two aromatic products [55] and [56]. The final structure of veatchine [54] surprisingly exhibits an absolute stereochemistry opposite to that of the steroids.

Wiesner's studies also led him to reinterpret Jacob's earlier work on the isomeric and chemically similar alkaloid atisine in terms of the formula [58], the six-membered ring for the alcohol being required by the IR spectrum of the corresponding ketone. Veatchine and atisine have each been converted by Pelletier to the same compound [57],

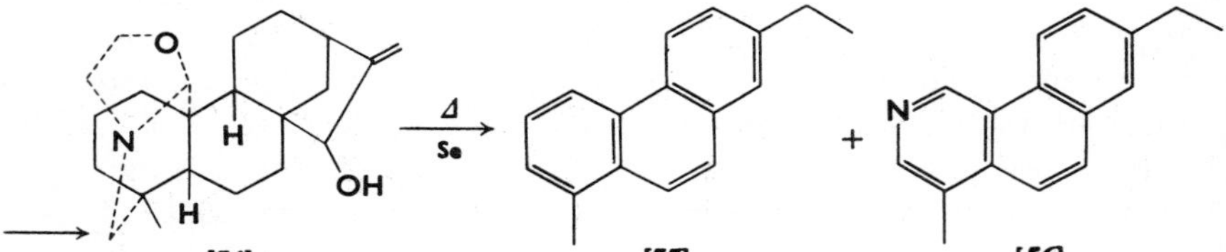

[54]
Veatchine

[55]   [56]

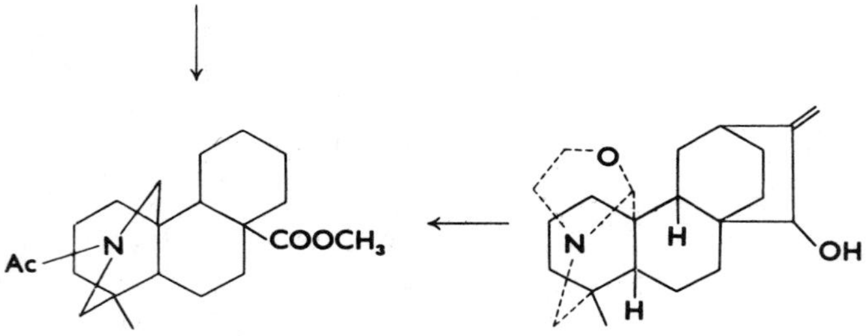

[57]   [58]
          Atisine

which retains all the original stereochemistry and so confirms the atisine formula. The reaction sequence on each alkaloid started from the imine (instead of the oxazolidine) and proceeded in the sequence: $NaBH_4$, $(CH_3CO)_2O$, $HO^\ominus$, $KMnO_4/HIO_4$, CH_2N_2, $HO^\ominus$, Br_2 (on $RCOO^\ominus Ag^\oplus$ salt), Zn/CH_3COOH.

4-7   Triterpenes

Triterpenes are the most widespread terpenes in plants, where they are almost invariably pentacyclic compounds with few functional groups and thus difficult to attack chemically for structural deduction. Accordingly, their structures were not known until after those of the biogenetically related steroids. The steroids and related C_{30} tetracyclic triterpenes, considered in Section 3–9, are almost always of animal origin. The pentacyclic triterpenes, from plants, afford the most characteristic picene dehydrogenation products only in meager yield if at all, and the major structure deductions, largely a product of Ruzicka's group, were based on the more common naphthalenic dehydrogenation products arising from cleavage in the center of the molecule (see Chart 1).

A potent tool for chemically interrelating the various members in order to fuse inferences made from studies on different compounds. was found in the reduction of –COOH to –CHO to –CH$_3$ so that the large group of triterpenes containing carboxyl at C-17 (cf. [2–77]) could be converted to the other large group bearing methyl there, as exemplified in the oleanolic acid [2–77] → β-amyrin [2–77] conversion. These triterpenes are always characterized by hydroxyl at the C-3 position as in steroids and commonly possess an exceedingly unreactive double bond at C-12. In the acids protonation of this bond leads to cyclization of a γ-lactone, protecting the double bond site so that oxidative incursions may be made in ring A or elsewhere, as with hot CrO$_3$ on the C-3 ketone and subsequent oxidative attrition in a series of steps as in the steroids (p. 129ff.). Characterization of the hydroxyl-*gem*-dimethyl moiety [59] is commonly made through the rearrangement to [60], initiated by PCl$_5$, and subsequent oxidation to acetone and a cyclopentanone (ring size/IR).

HO

[59]   [60]

The triterpenes arise biosynthetically by cyclization of a folded squalene chain, as discussed in Chapter 2. Moreover, the stereochemical configurations of the triterpenes are all predetermined by the way in which the squalene chain is folded! In an elegant elaboration of stereoelectronic possibilities, Ruzicka and his collaborators showed that all the known triterpenes could be derived *in vivo* from different spatial coilings of the acyclic squalene molecule prior to cyclization, followed by concerted 1,2-shifts of protons or methyl groups in stereoelectronically ideal *trans*-diaxial orientations. The simplest example may be found in the "all-chair" coiling of squalene shown in [61] which may be cyclized by HO$^\oplus$ (or its biochemical equivalent) to yield directly the C-3 alcohol ([62], X = OH) corresponding to hydroxyhopenone [63] in all stereochemical detail. Instead of hydrating, however, the carbonium ion resulting from the cyclization ([62],

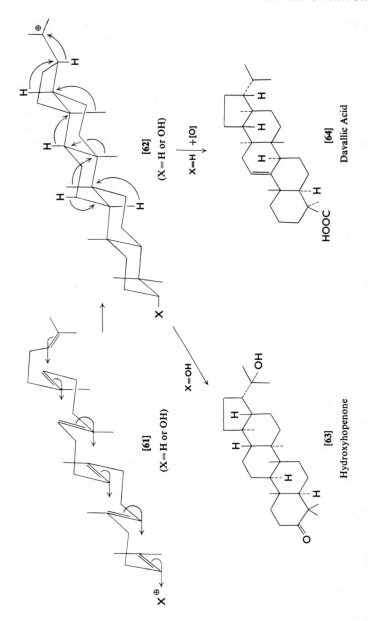

[62]
(X=H or OH)

X=H +[O]

[64]
Davallic Acid

HOOC

[61]
(X=H or OH)

X=OH

[63]
Hydroxyhopenone

X = H) may collapse by way of a series of concerted shifts of *trans*, antiparallel hydrogens and methyls as shown (arrows) for the formation of davallic acid [64]. The other triterpenes and steroids all arise similarly, via different initial conformations of squalene cyclization and methyl shifts.

A. AMYRIN FAMILY   The keystone of the structure proof of the β-amyrin triterpenes lies in an interesting and facile cleavage of the central ring which affords two more manageable decalin fragments. The acetate of oleanolic acid is cleaved to a diacid ([65], R = H) with concomitant lactonization when oxidized with CrO_3 and pyrolysis of the corresponding keto half-ester ([65], R = CH_3 and C-3 = ketone) smoothly yields the two compounds shown ([66] and [67]). These in turn yield 1,6- and 3,6-dimethylnaphthalene, respectively, on dehydrogenation, and [66] is interrelated with a previous degradation product of ambrein (p. 125). In another important series, SeO_2 oxidation of methyl acetyloleanate yielded a product [68] in which oxidation has "spread" from the olefin in every direction until blocked by quaternary centers. A similar oxidation was of great value in affording incursion into the central rings of the steroids (p. 134). Hydrolysis of the ester [68] then yielded an acid which readily decarboxylated (confirming its β-keto-acid nature); and the *cis*-1,4-orientation of the two introduced ketones was shown by formation of a pyridazine when treated with hydrazine (replacement of the two ketonic oxygens by $=N—N=$).

Corks being common in France, it is no surprise that they were extracted for natural products as early as 1807 by Chevreul, who isolated a wax since shown to contain largely the triterpene ketone, friedelin, which has the constitution [69]. An extraordinary rearrangement occurs when the derived olefin [70] is treated with strong acid, for the product is the previously known δ-amyrene [71]. This remarkable cascade of six migrating groups (see arrows) was a central feature of the proof of structure and stereochemistry, each migrating group being axial and hence *trans* and antiparallel for ideal stereoelectronic orientation. The reverse multiple migration is in fact a last step in the biosynthesis of friedelin!

B. ONOCERIN AND AMBREIN   Onocerin, an unusual plant triterpene elucidated by Barton, represents a triumph of biogenetic reasoning. The presence of only two hydroxy groups and two double bonds in

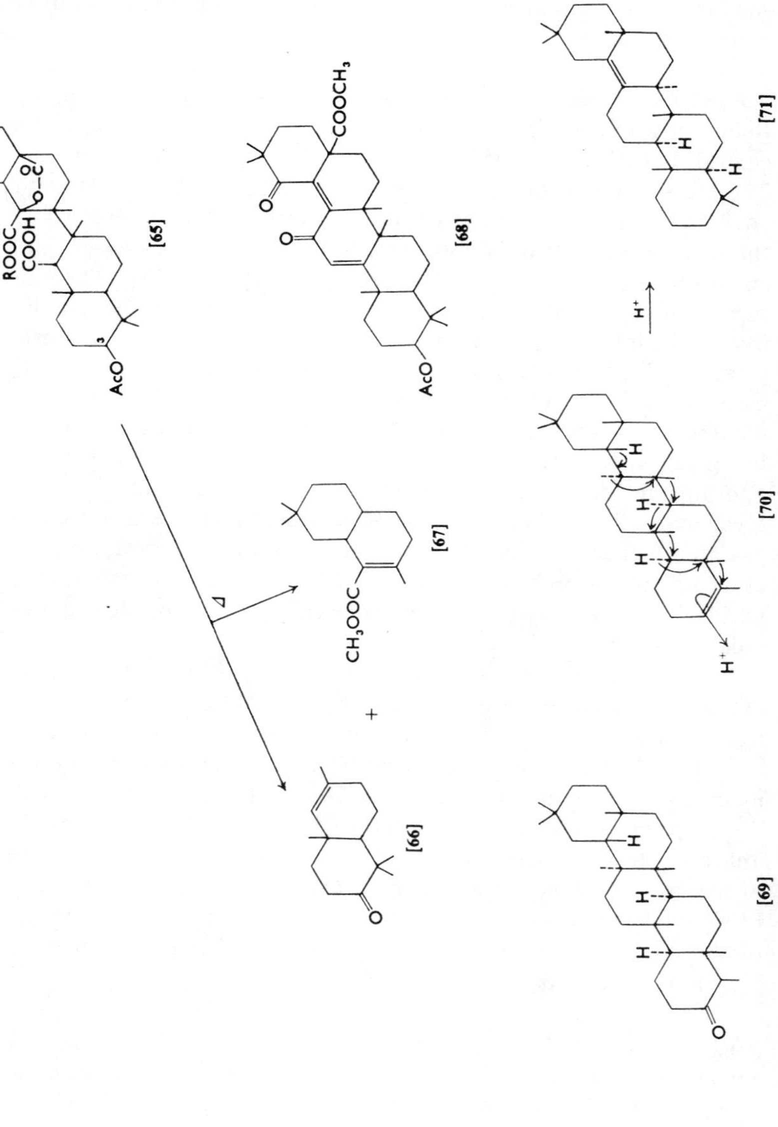

[65]

[68]

[67]

[66]

[71]

[70]

[69]
Friedelin

$C_{30}H_{50}O_2$ requires a tetracyclic structure and ozonolysis (of the diacetate) yielded formaldehyde and a diketone with two carbons missing, demonstrating the existence of two $C\!\!=\!\!CH_2$ groups. The diketone was carried through the PCl_5 rearrangement ([59]→[60]) followed by ozonolysis to two moles of acetone. These facts pointed to a symmetrical structure presumed to arise biosynthetically by squalene cyclization from both ends of the molecule and led to the formulation of onocerin as [72], conformation of which lent much support to the then new biogenetic cyclization mechanism for squalene.

In confirmation of the symmetry implied in [72]—the two halves are the *same*, not mirror images, so that the molecule *is* optically active— the equivalence of the two –OH groups was shown by the series of transformations, [73], in which each alcohol is separately oxidized to a ketone and reduced (W-K = Wolff-Kishner reduction) to methylene while the other is protected by acylation; each sequence yields the same compound. Also confirmatory is the isolation of only *three*, not *four*, tetrahydro-onocerins on catalytic hydrogenations.

Ambergris, the secretion of the sperm whale so highly valued in the perfume trade, is the source of a unique triterpene alcohol ($C_{30}H_{52}O$) called ambrein. That the molecule has two double bonds (and so must be tricyclic) is shown by its fission on ozonolysis into three molecules containing all the original carbons: CH_2O, a saturated diketone ($C_{12}H_{20}O_2$), and a saturated δ-lactone, ambreinolide ($C_{17}H_{28}O_2$). This cleavage into roughly equal parts much facilitated the structure work. The diketone readily underwent, in mild base, transformation to the unsaturated ketone [74], which was synthesized, and ambreinolide [75] was degraded by the Barbier-Wieland procedure to an acid previously known (and synthesized) from the diterpene manoöl [2–91]. (■) Ambrein is therefore [77], and ambreinolide was ingeniously synthesized by Lederer through a malonate alkylation with farnesyl bromide to give the acid [76] which cyclized to ambreinolide on treatment with formic acid, a reaction mimicking the biosynthetic one.

4–8   Carotenoids

The colors of carrots, tomatoes, and many other plants are largely caused by the presence of carotenoids, long-chain tetraterpene hydrocarbons. As a group they all have long sequences of conjugated double bonds, terminating at each end in a trimethylcyclohexene ring as in β-carotene [78], from carrots, or in compounds lacking these rings,

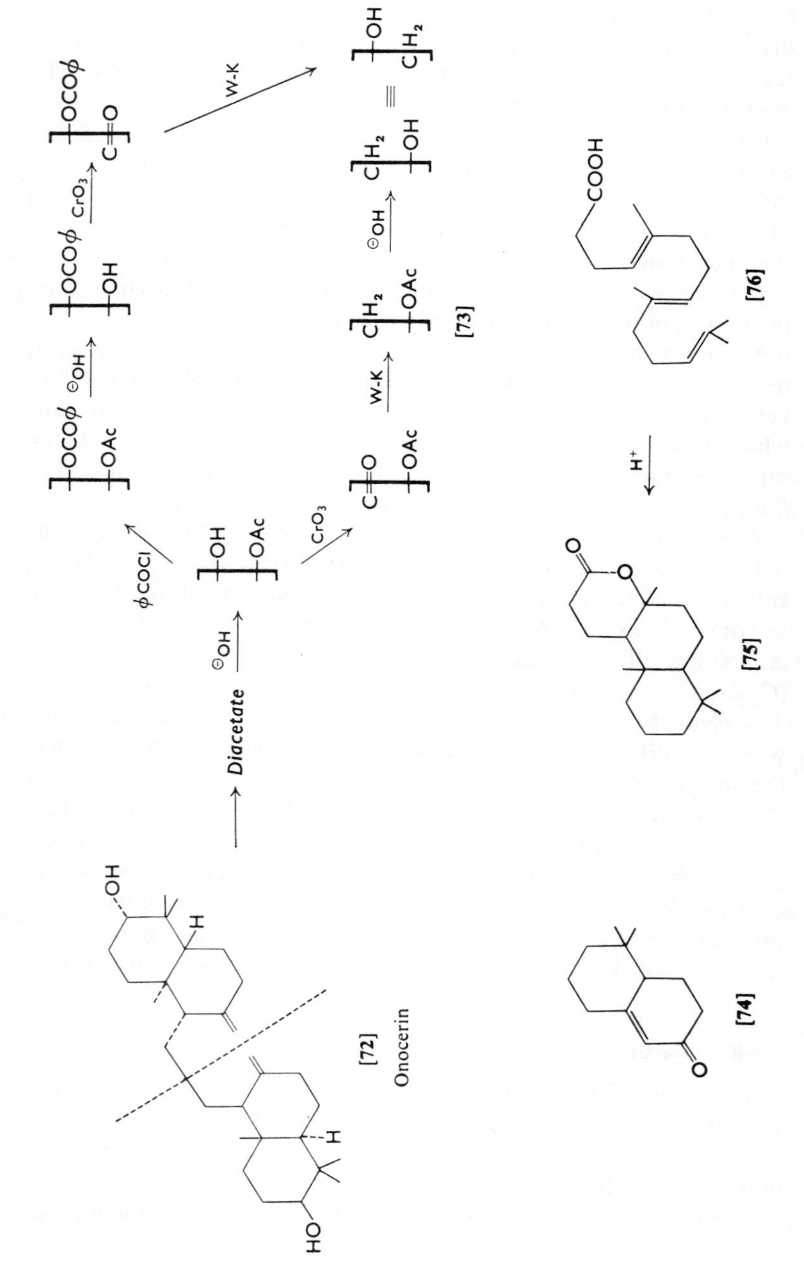

[72]
Onocerin

[73]

[74]

[75]

[76]

such as lycopene [79] from tomatoes. These pigments are converted in the human liver to vitamin A [80]. Hydrogenation of β-carotene ($C_{40}H_{56}$) yields a saturated $C_{40}H_{78}$ so that it contains eleven double bonds and two rings, and ozonolysis yields oxalic and acetic acids and geronic acid (2,2-dimethyl-6-ketoheptanoic acid). Taken with the isoprene rule these results suggest the structure [78] for β-carotene and this has been confirmed in an excellent synthesis by Isler. This process has in fact been worked out so successfully that it is currently used for the economic commercial production of various carotenes for food coloring.

Starting with synthetic β-ionone [81] the rationale involves adding one-, two-, and three-carbon units, the first one via a glycidic ester (Darzens reaction) which yields an unsaturated C_{14}-aldehyde on acid hydrolysis-decarboxylation. The subsequent chain-lengthening is accomplished by additions of R—CH=CH—OR' (R = H or CH$_3$) to the resultant aldehydes, a procedure which itself yields a new reactive aldehyde on hydrolysis. Two units of the C_{19}-aldehyde [82] so formed are then added to the *bis*-Grignard of acetylene to form a symmetric C_{40}-diol with a triple bond at the center. Double dehydration and partial hydrogenation of this triple bond complete the synthesis, which has also been successfully adapted to provide lycopene and several similar pigments.

4–9 Steroids

A. HISTORY AND BACKGROUND   The most extensive organic chemical studies on products of animal origin are unquestionably those on the steroids, dating back to the first decades of the last century with the isolation of crude preparations of cholesterol and cholic acid from bile and gallstones. Early in the twentieth century massive efforts were launched in the laboratories of Windaus, Wieland, and Diels aimed at deciphering the structures of these primary compounds, an enormously complex problem at that period, and not completed until 1932.

Once the basic skeleton was known, however, the structures of new steroids followed rapidly, the field picking up new impetus from the discovery in the late 1920's of the steroidal sex hormones. By World War II, most of the important steroid types were known, and attention turned increasingly to stereochemical elucidation. During the late 1940's, the discovery of cortisone and its broad medical efficacy lent a second huge impetus to steroid research, which turned to synthetic work aimed at producing and varying these cortical hormones.

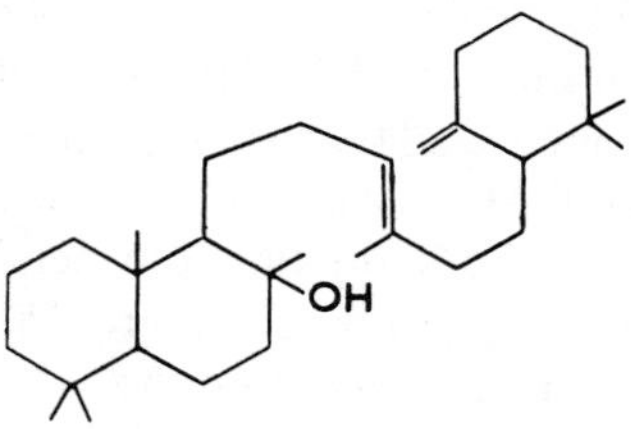

[77]

Ambrein

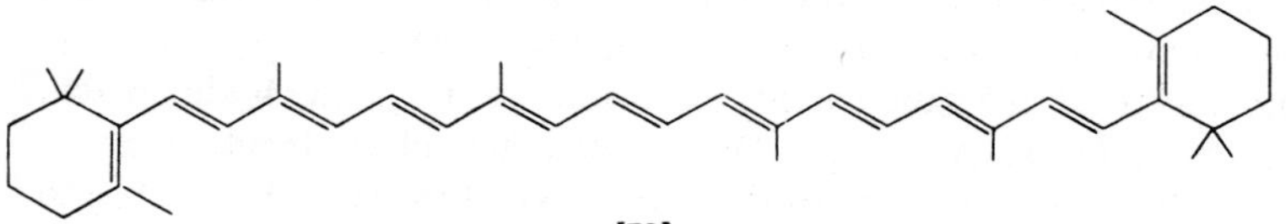

[78]

β-Carotene

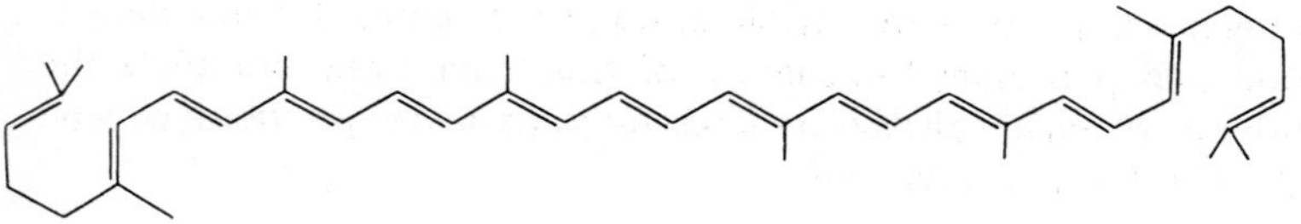

[79]

Lycopene

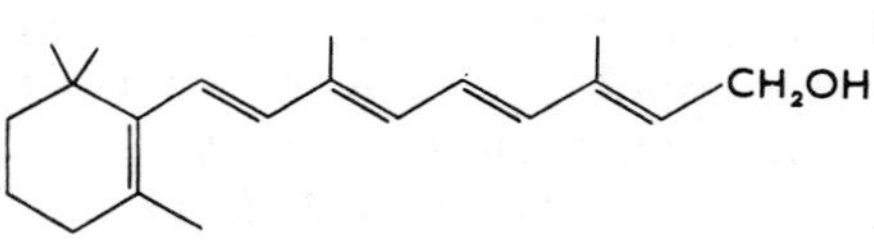

[80]

Vitamin A

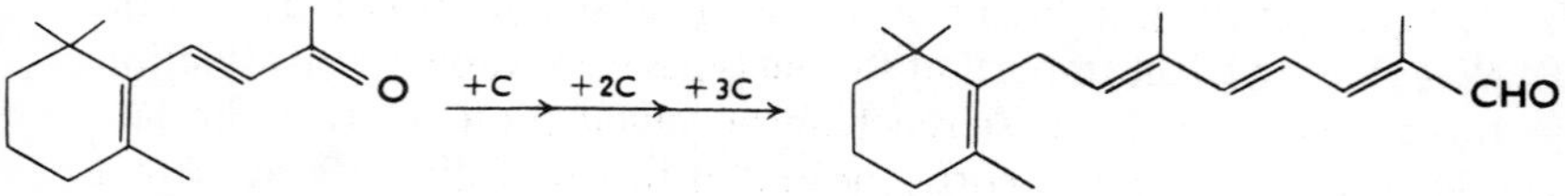

[81]

β-Ionone

[82]

Apart from these pharmaceutically oriented researches much independent study was concurrently being expended on structural studies of several other steroidal groups in plants: sapogenins, cardiac glycosides, and steroidal alkaloids. In recent years attention has increasingly focused on the biosynthesis and metabolism of these physiologically active molecules.

A notable function of the steroid skeleton for several decades has been its uniquely useful service as a rigid template for the development of new synthetic reactions, the correlation of physical measurements with molecular structure, and the testing of theoretical concepts. Thus, in large part the predictive correlations of UV spectra with unsaturated groups (cf. Woodward's Rules) were worked out on steroid examples, and Barton's initial elucidation of conformational analysis was supported almost exclusively by data from the enormous steroid literature. This literature continues to grow so prodigiously that the journal, *Steroids*, is now completely devoted to it.

In 1903 when Windaus began the steroid investigations that were to dominate his lifetime the empirical formula of cholesterol ($C_{27}H_{46}O$) was still regarded as uncertain. His early work was devoted to showing the β,γ-relation of the single double bond to the hydroxyl group. Diels began about the same time when he proved that the –OH was mounted on a ring by oxidation of cholesterol [2–94] to an unsaturated diacid with the same number of carbon atoms ($C_{27}H_{44}O_4$).

Wieland's concurrent studies on cholic acid [83] led to the formation of the reduced cholanic acid [83]. Windaus in 1919 fused these two lines of steroid investigation by preparing cholanic acid from coprostane (the desoxy-dihydrocholesterol with the A/B-*cis* ring fusion) by oxidation of the side chain. The oxidation also yielded acetone and similar oxidations on cholesteryl derivatives had previously yielded methyl isohexyl ketone, so that the nature of the side chain was also thereby clarified. The latter oxidations, although proceeding in very poor yield, afforded the important C_{19} derivatives (without side chain) which were later (p. 132) to be important in the sex hormone investigations.

The four rings of the steroid skeleton were attacked chiefly by oxidative cleavages to polyacids, and the use of the now notorious Blanc rule, which stated that pyrolysis of the substituted glutaric acid derived by oxidative cleavage of a substituted cyclopentanone will yield an anhydride, while the adipic acid from a cyclohexanone will be transformed to a five-membered cyclic ketone. This is exemplified in

Wieland's conversion of desoxycholic acid [83] to desoxybilianic acid [84] and its subsequent pyrolysis to a ketone which implies a six-membered ring A. Similarly, oxidative attrition of the cholanic acid side chain via the Barbier-Wieland degradation led to etiocholanic acid [85], oxidation of which afforded the diacid [88], which formed an anhydride on pyrolysis and led to assumption of a five-membered ring D. The rule failed, however, in the experiments from which the size of the C-ring was inferred and in 1928, when Wieland and Windaus received Nobel Prizes for their studies, their formula for cholesterol was [86].

The first instance of the use of X-ray crystallography to provide new structural information in an organic problem came with Bernal's announcement in 1932 that the ergosterol molecule had a flat, lath-like shape and so could not have the more globular form implicit in [86]. Furthermore, a few years earlier, Diels had obtained chrysene by dehydrogenating cholesterol. These important clues very shortly led to the presentation of the correct structure of cholesterol [2–94]. From this hard-won achievement, steroid work entered a new phase in which many structures were assigned much more rapidly, first by formation of Diels' hydrocarbon, then by chemical interrelation of the new steroids with cholesterol and the cholic acids.

The first sex hormone to be isolated, from pregnancy urine, was estrone (1929), shown to have a ketone (via oxime) and phenolic hydroxyl and no unsaturation beyond that of the aromatic phenol ring, as shown by hydrogenation studies. Its formula ($C_{18}H_{22}O_2$) thus requires that estrone be tetracyclic, establishing an inference of relation to the steroids. After removal of the ketone group in estrone methyl ether by Wolff-Kishner reduction, selenium dehydrogenation yielded a tetracyclic phenanthrene shown synthetically by Cook to be the desmethyl derivative of the Diels hydrocarbon. Cook also demonstrated the ketone position as C-17 by initial treatment of estrone methyl ether with methyl magnesium bromide to provide a marker before dehydrogenation. Thus a family of steroids without the side chain became evident. In synthetic attempts to raise the medical potency of estrone, its C-17 dihydro derivative, estradiol [2–101], was soon found to be identical with the primary active hormone which had been isolated with great difficulty from the ovarian tissue of 50,000 sows.

An almost identical history applies to the male hormones: the less active and secondary, excreted hormone, androsterone, was isolated

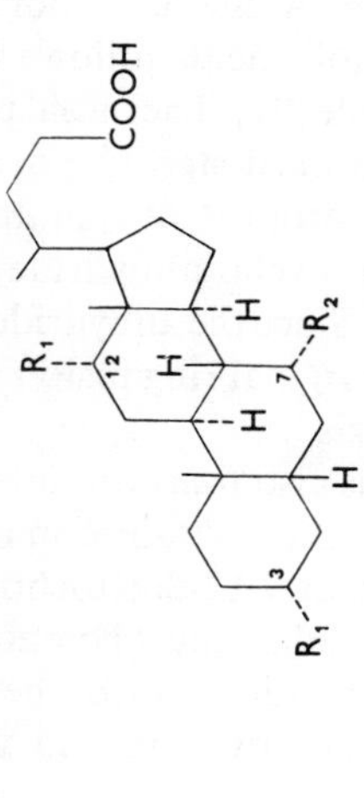

[85]

[88]

[84]

[87]

Androsterone

[83]

Cholic Acid: $R_1 = R_2 = OH$
Desoxycholic Acid: $R_1 = OH; R_2 = H$
Cholanic Acid: $R_1 = R_2 = H$

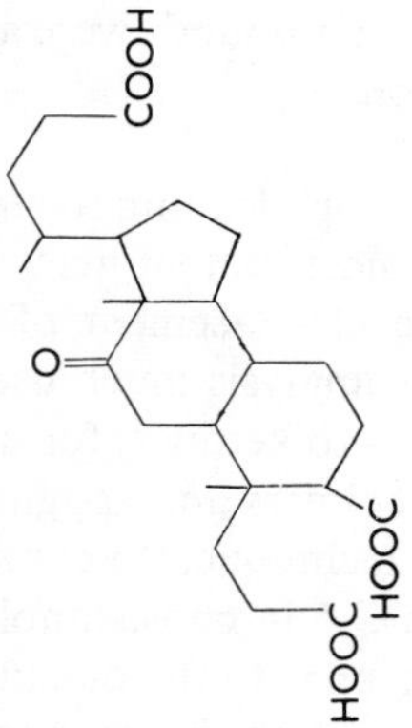

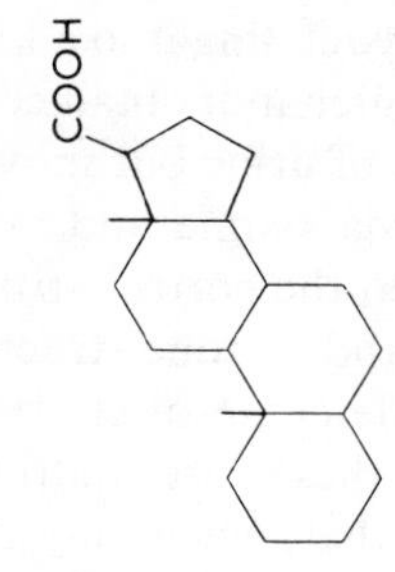

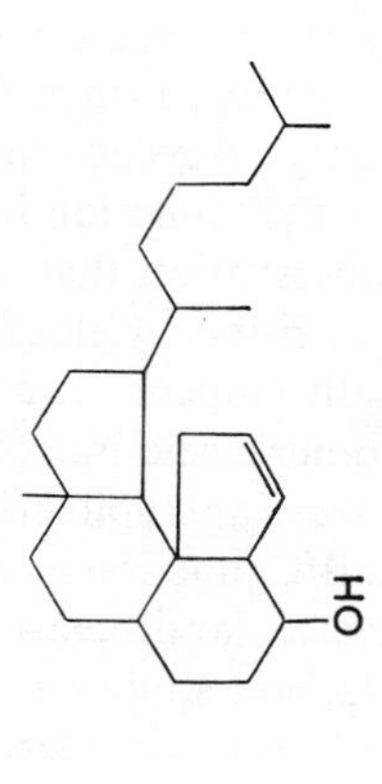

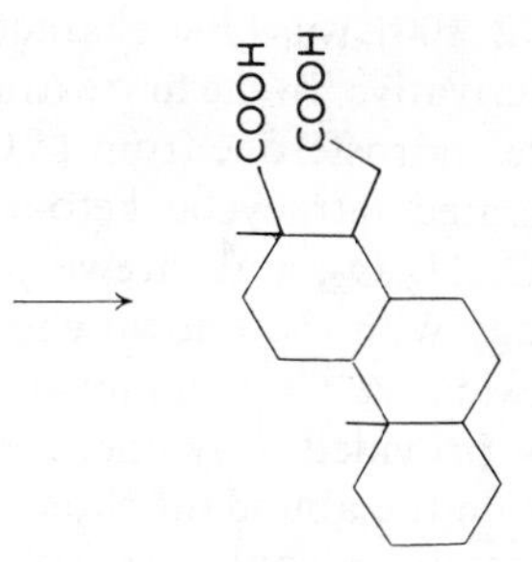

[86]

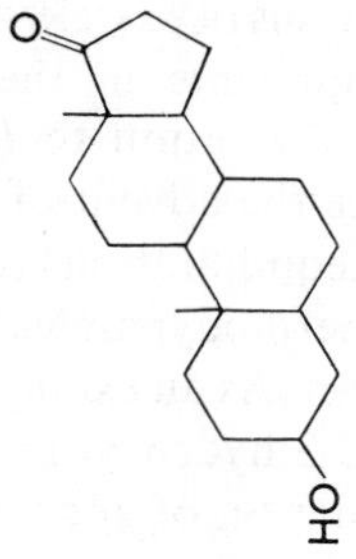

from male urine and characterized first. The primary hormone, testosterone [2–100], was later characterized by relating it to a known androstane derivative. In the first extraction, Butenandt obtained only 15 mg of pure androsterone from 15,000 liters of urine but showed it to be a saturated tetracyclic keto-alcohol (via oxime and acetate formation), $C_{19}H_{30}O_2$, and shrewdly suggested the correct structure [87] by analogy with the known cholesterol and estrone structures! The earlier oxidative CrO_3 removal of the side chain of cholesterol (p. 129) now provided a synthetic route to these rare androstane derivatives from the abundant cholesterol, so that, when only 10 mg of pure testosterone could be isolated as the hormonally active constituent of 100 kg of steer testicular tissue, compounds were available for determining its identity by comparison.

B. STEREOCHEMISTRY: GENERAL   In earlier work leading to stereochemical assignments in the steroids the most potent technique involved ring formation to demonstrate the *cis*-placement of substituents. Since the advent of conformational analysis much use has been made of equilibrations (as of hydrogen α- to ketones) for determination of thermodynamically more stable epimers and assignment of configuration. As an example of the earlier technique, the configuration of the C-3 hydroxyl relative to that at C-5 in cholestanol was shown by oxidation of appropriate ring-B ketones to the diacid [89] which forms a lactone acid on treatment with acetic anhydride and so must possess *cis*-COOH and –OH groups. This behavior is of course not shown by either the C-3 or C-5 epimers. A similar proof of a *cis*-junction between rings A and B in the cholic acids follows from the base-catalyzed cyclization of [90] to the oxide [91]. The initial proof of the *trans*-C/D junction in cholic acid (and most steroids) derived from the observation that the triacid [92], a product of a multistep oxidative attrition, yielded on pyrolysis a cyclic anhydride that hydrolytically reopened to an isomeric triacid. Since the anhydride and thus the isomeric acid must be *cis*-fused, the original ring fusion must have been *trans* and epimerized in the pyrolysis.

That the B/C junction in common steroids is also *trans* was inferred from the molecular dimension of about 4 × 7 × 20 Å obtained in many X-ray diffraction studies as well as the fact that 7,12-diketocholanic acid (see [83]) is not isomerized by hot alkali and so must contain the most stable orientation at the B/C-junction adjacent to the C-7 ketone grouping. Thus many observations of this kind and inter-

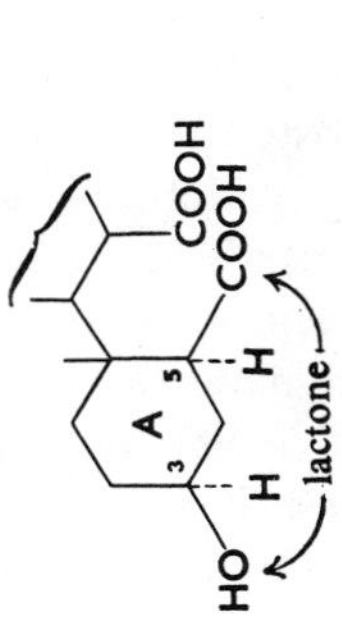
HO
H
lactone
3
A
5
H
COOH
COOH

[89]

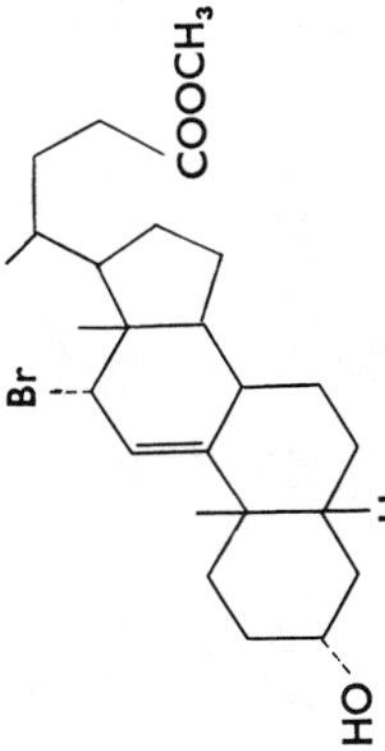
HO
H
Br
COOCH₃

[90]

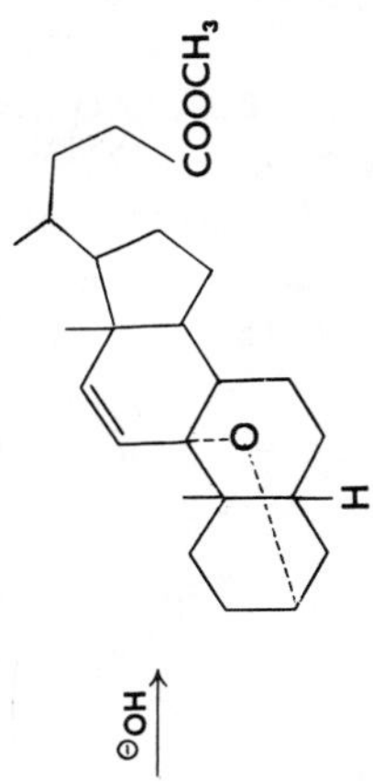
⊖OH
O
H
COOCH₃

[91]

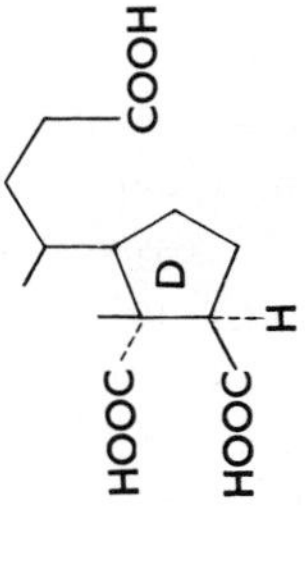
HOOC
HOOC
D
H
COOH

[92]

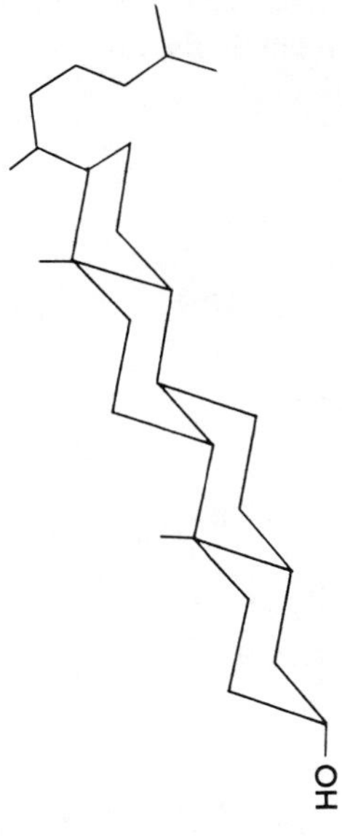
HO

[93]

relations of different steroids have served to elucidate fully the many stereochemical details in the various examples. Almost all steroids have the absolute stereochemistry embodied in the projection [93] for cholestanol. In steroidal nomenclature, substituents below the molecular plane are designated α, those above, β; cholestanol is thus 3β-cholestanol and has the 5α-hydrogen, whereas the A/B-*cis* isomer coprostanol is 5β-H, as are the bile acids (cf. [83]).

C. STEREOCHEMISTRY: TIRUCALLOL   The stereochemistry of tirucallol [94] was solved by Jeger and co-workers via a very elegant interrelation

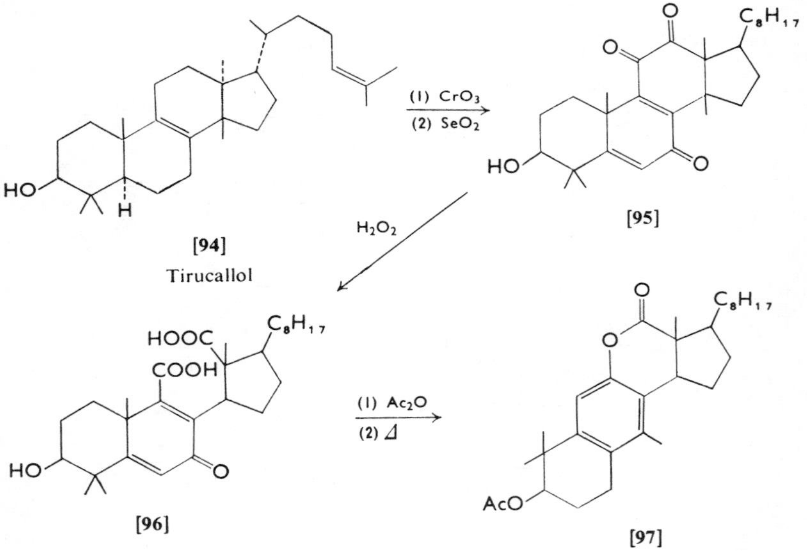

[94]
Tirucallol

(1) CrO₃
(2) SeO₂

[95]

H₂O₂

[96]

(1) Ac₂O
(2) Δ

[97]

with the isomeric lanosterol, from which it differed only in configuration; but the total stereochemistry of lanosterol was known. Each was submitted to successive CrO_3 and SeO_2 oxidations leading to [95] (stereochemistry omitted), an important intermediate earlier in establishing the environs of the B/C-ring system of lanosterol. This was oxidized by H_2O_2 to the diacid [96], which had been shown by Barton to undergo an intriguing rearrangement to [97] on acetylation and pyrolysis. The lactones [97] obtained from lanosterol and tirucallol (the 3-epimer of tirucallol was actually used since all the natural steroids have the 3β-OH configuration) were then found to be identical in every respect except that they had equal and opposite optical rotations, thus proving that they possessed opposite configurations at *every* asymmetric center in the lactone and therefore at C-13, C-14, C-17, and C-20 in the natural compounds. This completely defined the stereochemistry of tirucallol.

D. STEROIDAL ALKALOIDS   Cholesterol is $C_{27}H_{46}O$; solanidine (from potato plants) is $C_{27}H_{43}NO$; both are transformed into Diels' hydrocarbon on selenium dehydrogenation. Replacement of three hydrogens in cholesterol by nitrogen could yield a solanidine structure since solanidine is a tertiary amine as well as an unsaturated alcohol. The placement of the nitrogen in fact follows from the concomitant isolation of 2-ethyl-5-methyl-pyridine in the dehydrogenation. This pyridine is also obtained from most other steroid alkaloids. Hence the simplest formula for solanidine would be [98], and this has been confirmed by interconversions with sapogenins.

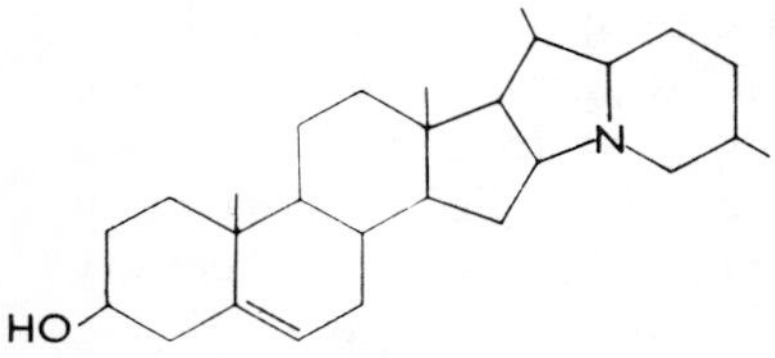

[98]

Solanidine

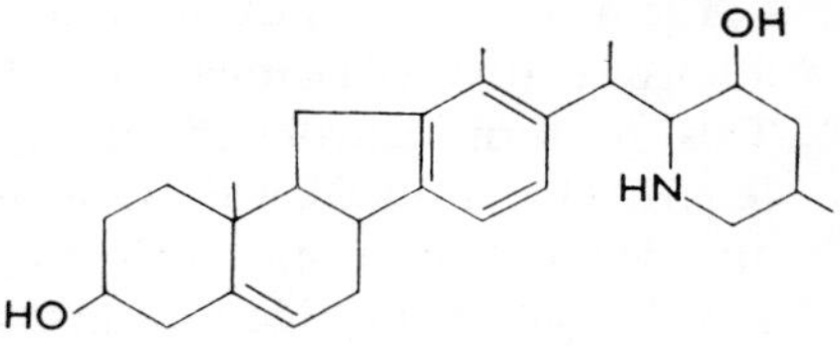

[99]

Veratramine

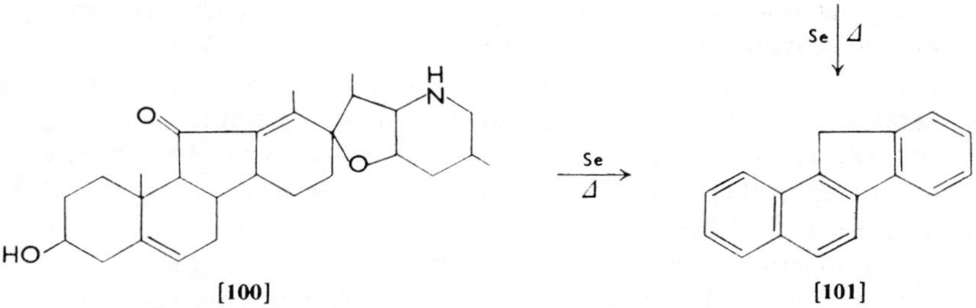

[100]

Jervine

[101]

The *Veratrum* alkaloids, however, yield fluorenes on dehydrogenation (as well as 2-ethyl-5-methyl-pyridine), as in [101] obtained from veratramine [99] and jervine [100]. This observation led Wintersteiner to propose a rearranged steroid skeleton with a five-membered

C-ring. Solvolysis of a normal steroid 12β-alcohol tosylate has also afforded this same skeleton since the 12β-group bears the stereoelectronically correct parallel orientation for rearrangement of the C/D-ring juncture bond.

Cevine ($C_{27}H_{43}NO_8$) and the isomeric germine are more complex in having many more hydroxyls. The steroidal skeleton is clear, however, from the formation of the dehydrogenation product [102] which could be hydrogenated to a basic octahydro derivative recognized as a fluorene with an unconjugated amino group from its UV spectrum. The isolation of 2-ethyl-5-methyl-pyridine and [102] allowed the skeleton ([102] + C—N bond at arrow) to be written for cevine very early in the work and the problem then centered on placing the oxygen functions on this frame. As cevine has no olefinic or ketonic groups and all the carbons are accounted for in [102], consideration of the empirical formula shows that the eight oxygens of cevine can only be seven hydroxyls and one ether link; the ether was accounted for by Barton as a cyclic hemiketal.

The major clue to placement of these groups came from the formation of a saturated lactone triacid ($C_{14}H_{18}O_8$) on chromate oxidation. This in turn dehydrated on pyrolysis to form decevinic acid ($C_{14}H_{14}O_6$), a diacid with an enolic grouping ($FeCl_3$ color) which afforded [104] on selenium dehydrogenation. Decevinic acid chemistry posed an interesting riddle (cf. alkali yielded a dibasic acid, $C_{13}H_{16}O_5$, which changed to a saturated ketone-γ-lactone, $C_{12}H_{16}O_3$, on heating), Woodward's decipherment of which in terms of the structure [103] opened the way for placement of the hydroxyls in cevine.

The lactone triacid was assigned structure [105] which then could only arise from ring B, via oxidative cleavage of rings A and C. This permits specific placement of four of the oxygens (including the hemiketal) in these rings. Reasoning from the quantitative uptake of HIO_4 in a series of cevine derivatives with the hydroxyls increasingly acetylated further allowed exclusion of a number of sites for hydroxyl attachment in the regions of the previously fixed hydroxyls and the amino function. The cevine structure finally deduced is [106].

E. SYNTHESES Partial syntheses in the steroid field have become an intensely developed area owing to the need to prepare the rare hormonal steroids from common steroids and so make them medically available. The sophistication, length (often twenty successive reactions), and high yields of these syntheses are an impressive tribute to the power of contemporary synthetic chemistry. A cheap source of available steroids in quantity often is the sapogenins found in various plants. A major route for their conversion to hormonal steroids with the two-carbon side chain was first developed by Marker and is illustrated in [107]→[109] for the sapogenin, diosgenin [107]. Conversion of [107] to the enol ether [108] allows normal oxidative

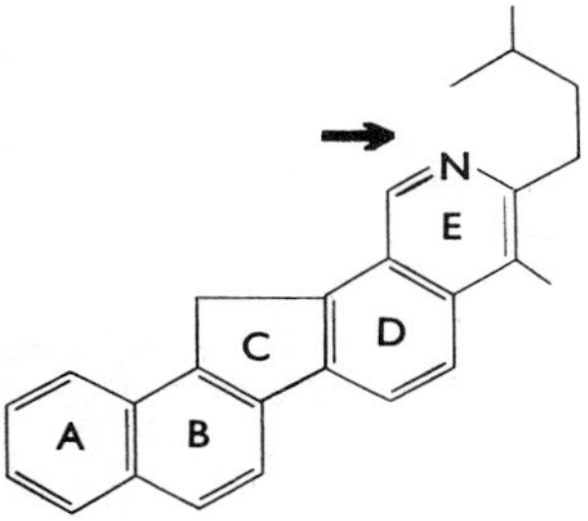

[102] 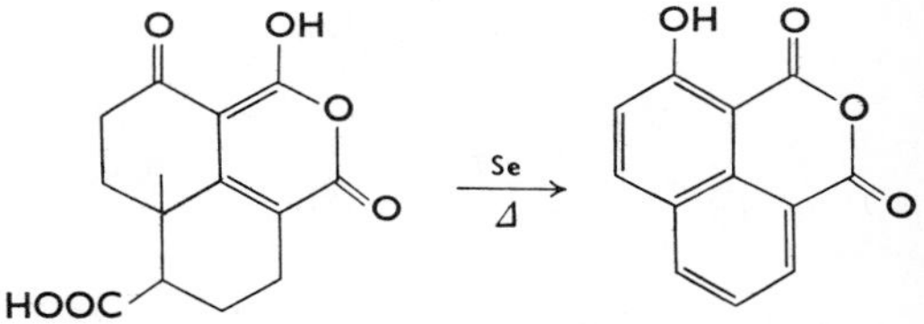   [103] $\xrightarrow{\underset{\Delta}{Se}}$ [104]

HOOC

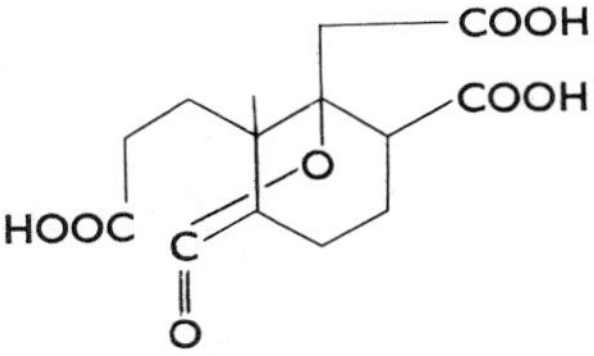

[105]

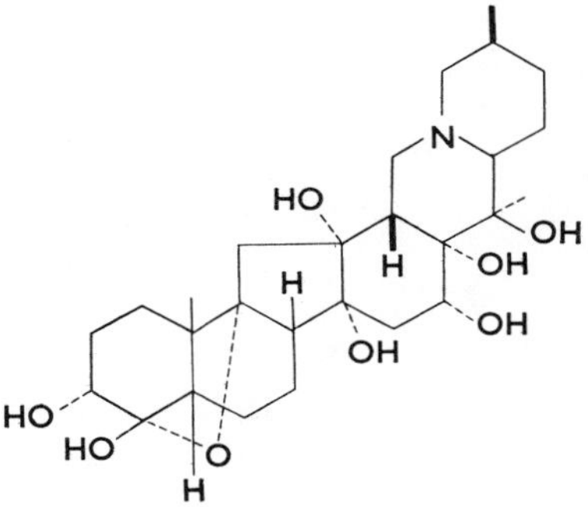

[106]
Cevine

cleavage to a keto-ester followed by β-elimination to [109]. An auxiliary boon for partial synthesis of the cortical steroids bearing the otherwise uncommon 11β-hydroxyl appeared with the announcement by the research group at the Upjohn Company of a microorganism which metabolizes certain steroids solely by hydroxylation at C-11. This microbiological oxidation, in which the steroid is "fed" to mold cultures and later reisolated, has since become a potent synthetic tool for introducing substituents into a variety of unactivated positions on the steroid nucleus. The mechanisms of these reactions will surely constitute an interesting chapter in the future of organic chemistry.

Total syntheses of the nonaromatic steroid skeleton, which tied into

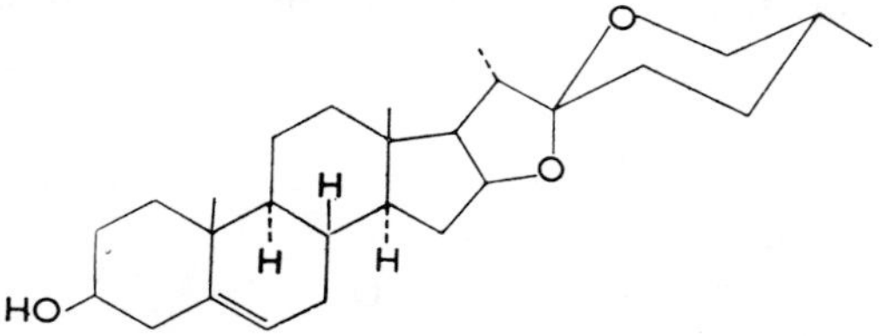

[107]

Diosgenin

Ac₂O

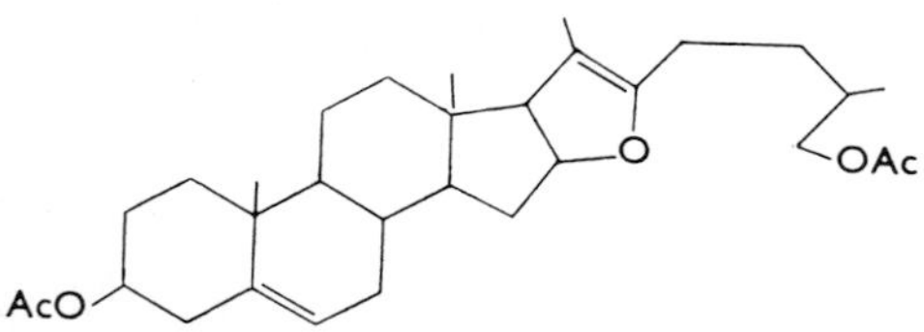

[108]

(1) CrO₃ | (2) HOAc

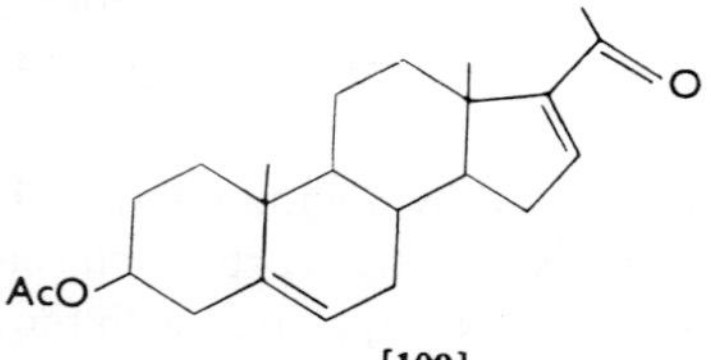

[109]

previous partial syntheses of most of the common examples, were first independently announced in 1951 by Woodward and by Robinson. The more flexible Woodward synthesis (which included a double bond at C-11 for conversion to cortical hormones) was subsequently

CHART 2 ■

Total Synthesis of Epiandrosterone

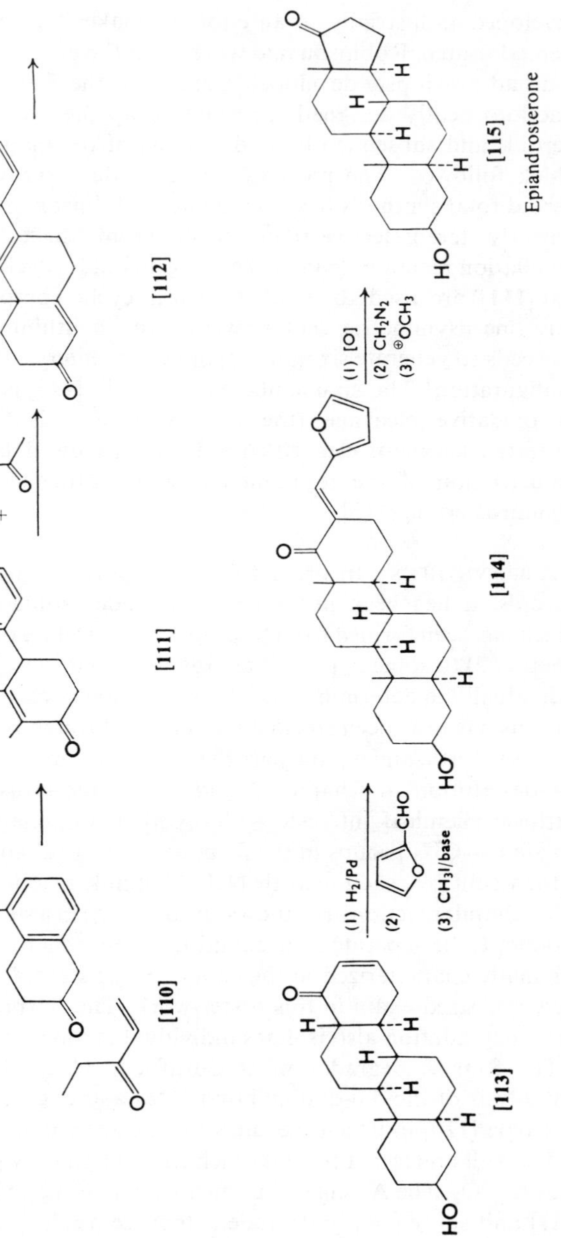
OCH₃
[110]
OCH₃
[111]
OCH₃
[112]
[113]
(1) H₂/Pd
(2) CHO
(3) CH₃I/base
HO
[114]
(1) [O]
(2) CH₂N₂
(3) ⊕OCH₃
HO
[115]
Epiandrosterone

140   ■ THE MOLECULES OF NATURE

developed industrially in an effort to make it a competitive commercial source. Robinson had worked on the problem for many years and had previously developed a reaction (the Robinson annellation reaction) neatly designed for building up the cyclohexane rings of steroids and subsequently used in most of the many total syntheses which followed. The most extensive modern study of methods of steroid total synthesis has been made by Johnson and resulted in the masterly stereoselective sequence shown in Chart 2. Two Robinson annellation reactions (Michael additions + aldol cyclization; see [110] and [111]) are used to build up a tetracyclic compound [112] with only one asymmetric center, whereupon a lithium reduction then proceeds to generate six more asymmetric centers, all with the correct configuration! The six-membered ring D in [113] is then contracted by oxidative cleavage (the furfurylidene derivative [114] of the saturated ketone is cleaved to a diacid) followed by a Dieckmann condensation of the resultant diester to form the cyclopentanone, epiandrosterone [115].

F. BIOSYNTHESIS   In order to find support for the biosynthesis of steroids, it has been necessary to degrade radioactive cholesterol, which has been formed from feeding labeled acetate or mevalonate (cf. Chapter 2) to some appropriate organism, and to isolate each carbon individually to determine its C^{14}-content. Every carbon of the steroid nucleus has now been so isolated, largely through the work of Bloch and Cornforth, and not only are the results in complete agreement with the description in Chapter 2, but the methods used are often of intrinsic chemical interest. Kuhn-Roth determinations, of course, isolate $C—CH_3$ groups in the form of acetic acid, and the acetic acid is then split by reaction with NH_3 (Schmidt reaction; see Chart 3). The Schmidt reaction also allows any carboxyl in a steroid degradation product to be separated for counting, so that many of the polyacids originally characterized in the course of the steroid structure studies have appeared again in this tracer work. The Barbier-Wieland sidechain degradation also isolates individual carbons for counting.

    The Ring A degradation of Cornforth (Chart 3) is founded on ozonolysis of the 5,6-double bond [116] and degradation to a ketoaldehyde [120] in which the carbonyl groups are in a 1,5-orientation and so will undergo a reverse Michael reaction on vigorous pyrolysis with K_2CO_3. The A-ring is thus distilled out as methyl-cyclohexanone [121] and subsequently degraded stepwise as shown in the chart to

Degradation of Steroid Ring A in Tracer Studies

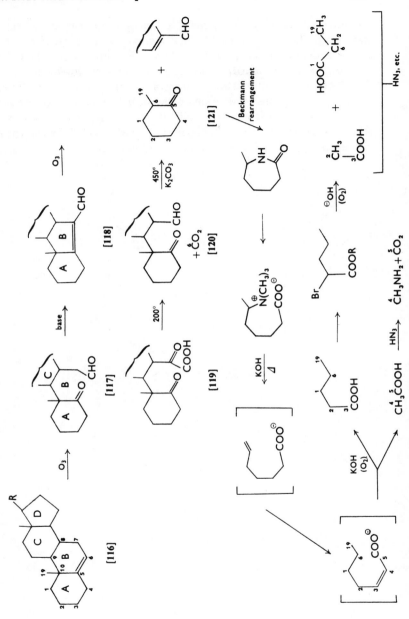

isolate each carbon (numbered with their steroidal ring positions) for counting. The same reverse Michael pyrolysis was utilized in ring D on the ozonolysis product [122] from the 14,15-double bond. The volatile aldehyde distilled out here [123] was used not only for tracer degradation but also in establishing the absolute configuration at C-20 (and from it, by previous *relative* stereochemical correlations, the absolute configuration of the whole steroid skeleton). This was possible since the dihydro derivative of [123] could be synthesized from the terpene citronellal [124], of known absolute configuration.

[122]   [123]

[124]

$$5$$

■ ALKALOIDS

5–1   Alicyclic Alkaloids

The chemistry of alkaloids has always been an active and important part of organic chemistry and has provided many of the most challenging and subtle of all chemical problems, both of deduction and of synthesis. Nevertheless, not all alkaloids are complex. A great many relatively simple natural amines were isolated and elucidated early in the active history of alkaloid chemistry (late nineteenth century) and provided a testing ground for many of the degradative tools which were later used on more complex molecules. Thus coniine, the major alkaloid of Socrates' poison cup, is $C_8H_{17}N$ and saturated; hence it has one ring. CrO_3 oxidation yielded n-butyric acid while AgOAc or zinc dust distillation yielded a pyridine, $C_8H_{11}N$, which was identified via its oxidation to pyridine-2-carboxylic acid. (■) Coniine is therefore 2-n-propylpiperidine, and the deduction was verified by synthesis.

A. LUPIN ALKALOIDS   Slightly more complex is the bicyclic alkaloid lupinine ($C_{10}H_{19}NO$) from the seeds of the yellow lupin. The presence of an alcohol grouping was shown by the preparation of derivatives (with $\phi COCl$, ϕNCO) and dehydration (with H_2SO_4) to an anhydro compound. Oxidation to lupininic acid ($C_{10}H_{17}NO_2$) showed the alcohol to be primary. The classical Hofmann degradation, so common in alkaloidal structural work, was applied three times to lupinine before trimethylamine and a nitrogen-free product were obtained, thus demonstrating the presence of a tertiary amine embedded in a ring system.

The first and second Hofmann degradations were each followed by hydrogenation of the olefin formed, but the third yielded a methine containing a vinyl group as was shown by its oxidation ($KMnO_4$) to HCOOH and an acid ($C_9H_{18}O_3$) which lactonized to the hydroxyl retained throughout the degradation from lupinine. The third methine was also hydrogenated and its hydroxyl was converted (via HBr, then $(CH_3)_3N$) to a quaternary ammonium salt. The salt still retained optical activity, indicating that at least one asymmetric center in lupinine had remained undestroyed to this point, but Hofmann degradation of this derivative yielded an acyclic olefin ($C_{10}H_{20}$) with no optical activity. Since this olefin could also be oxidized ($KMnO_4$) to HCOOH, and a ketone ($C_9H_{18}O$), the grouping

$$\underset{\underset{*}{C}-\overset{\displaystyle \overset{\textstyle CH_2}{\|}}{C}-C}{}$$

must be present, the asterisked carbon having been the asymmetric one in the previous compounds.

The inference that the ketone is 4-nonanone derives from a Beckmann rearrangement on its oxime which affords *n*-amyl *n*-butyramide. From this series of reactions one may construct only six structures for lupinine (at a later time the Kuhn-Roth determination would have eliminated four of these which bear C—CH$_3$ groups) if no three- and four-membered rings are accepted. Of these Willstätter selected [1] for lupinine and this has been confirmed by other degradative studies and by synthesis.

Amongst the other lupin alkaloids, cytisine [2], the toxic principle of laburnum, was shown to be a secondary amine and only two Hofmann degradations sufficed to remove the basic nitrogen. The methines were hydrogenated at each stage and after the second stage, ozonolysis of the pyridone ring yielded a lactam [3] which gave α,α'-dimethyl-glutaric acid on oxidation. As the chemical similarity of anagyrine

[4] was soon recognized, it was subjected to a similar degradation, resulting in α-methyl-α'-*n*-amyl-glutaric acid.

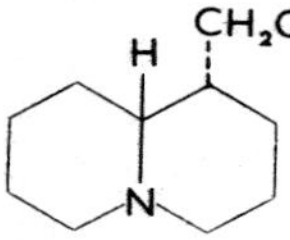

[1]

Lupinine

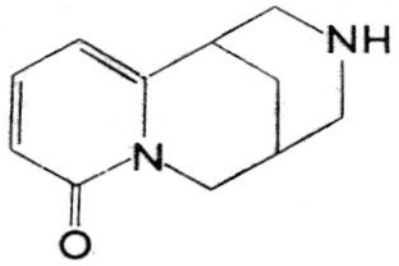

[2]

Cytisine

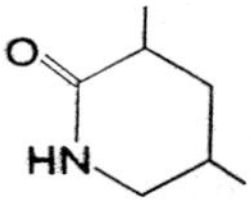

[3]

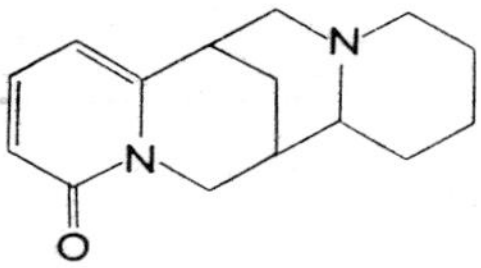

[4]

Anagyrine

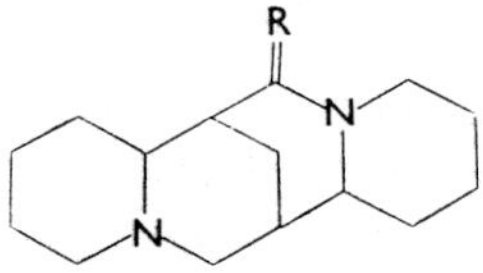

[5]

Sparteine, R = H$_2$
Oxysparteine, R = O

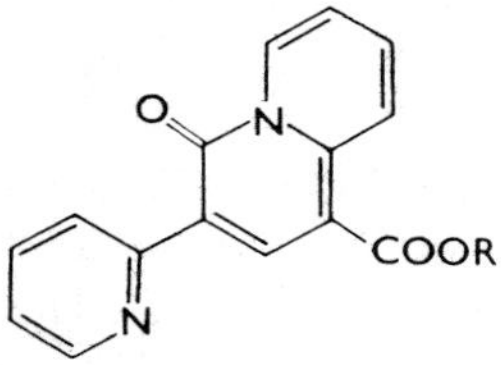

[6]

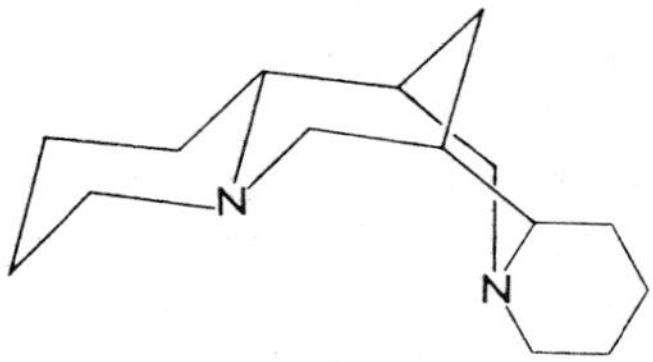

[7]

Sparteine

Concurrently, work on sparteine **[5]** had proceeded through *six* successive Hofmann degradations, demonstrating two ring-locked tertiary nitrogens but little else except HCOOH on oxidation of the final non-nitrogenous C$_{15}$-hexaene. When each of the Hofmann degradations was followed by hydrogenation, a saturated C$_{15}$ hydrocarbon was ultimately obtained but was too similar to several known isomers to distinguish it by the physical properties available for study in the 1930's. The structure was deduced following correlation of the anagyrine and sparteine work, achieved by reduction of anagyrine to sparteine. Oxysparteine [5] was synthesized by Clemo (1936) via condensation of ethyl-2-pyridyl-acetate with ethyl ortho-formate and acetic anhydride to [6] followed by hydrogenation and ester reduction (Na/ROH) to a primary alcohol, which was subse-, quently cyclized to oxysparteine by conversion to the bromide with HBr and heating. When $LiAlH_4$ was later developed, it smoothly converted oxysparteine to sparteine. The absolute configuration of natural l-sparteine is shown in [7].

B. PYRROLIZIDINE ALKALOIDS   The many alkaloids of *Senecio* species are characterized by a group of a few "necine" bases or their N-oxides with hydroxyl groups esterified by C_5- or C_{10}-terpene acids, sometimes as cyclic dilactones, viz., junceine [2–116]. Early work by Menshikov on unraveling the necine structure problem, continued and completed by Adams in the 1930's, offers an example of the classical sequence of determination: nature of the functional groups, skeleton, location of the functional groups, and stereochemistry.

Retronecine ($C_8H_{13}NO_2$) is basic and unreactive to HNO_2, while the Zerewitinoff determination shows two hydroxyl groups. Raney nickel hydrogenation affords platynecine, a dihydro compound and itself a natural necine base from other *Senecio* alkaloids. Neither base shows any N—CH_3 or O—CH_3 (by Zeisel) so that retronecine is a bicyclic tertiary amine with one double bond and two alcohol func-tions, reductive removal of which led to the saturated bicyclic parent amine, heliotridane.

The skeleton of heliotridane was elucidated by Emde reductions, a variant on the Hofmann degradation in which a quaternary ammon-ium salt is reduced (Na/Hg) by cleavage of one C—$N^{\oplus}$ bond. The first Emde procedure resulted in a base which was easily dehydrogenated to a pyrrole and identified by its chemical properties. The second Emde yielded 3-methyl-4-dimethylamino-heptane (■), verified by synthesis. These results require the structure [8] for heliotridane.

The location of the functional groups in retronecine then follows from several further observations. Platinum-catalyzed hydrogenation of retronecine, but not platynecine, causes hydrogenolysis of one –OH, which is therefore allylic. Rate studies on esterification of the two alcohol groupings in several derivatives and models showed the allylic alcohol to be primary, the other one secondary. Also Kuhn-Roth determinations show C—CH_3 in the hydrogenolyzed alcohol but not in the natural alkaloids. The greater basicity of platynecine over retronecine is not consistent with the presence of an eneamine (C=C—N) so that the only location for the allylic primary alcohol moiety is that depicted in the retronecine structure [9]. The formation of a cyclic ether between the alcohol groupings suggests the position of the secondary alcohol as shown in [9]. That the stereochemistry is represented correctly in [9] follows from the necessary all-*cis* stereochemistry of the strained cyclic ether [10]; the correct absolute stereochemistry is shown. (±)-Retronecine has been synthesized by Geissman.

[8]   [9]   [10]

Retronecine

C. TROPANE ALKALOIDS The tropane alkaloids include cocaine and scopolamine, the "truth serum," as well as atropine, and other constituents of such plants as henbane, mandrake, and deadly nightshade—known even in medieval times for their hallucinogenic qualities. The alkaloids are generally esters of tropine or closely related bases which were intensively studied in the last century and resulted in a structure for and confirmatory synthesis of atropine by Willstätter at the turn of the century. Hyoscyamine ($C_{17}H_{23}NO_3$) is readily racemized to atropine under weakly basic conditions and both alkaloids yield tropic acid (α-hydroxymethyl-phenylacetic acid) and tropine ($C_8H_{15}NO$) on saponification. Other esters of tropic acid are racemized equally easily and tropine is optically inactive.

This important result, not fully appreciated in its time, requires that tropine be either a meso-compound or without asymmetric carbons.[1] This puts severe restrictions on the possible structures of tropine.

---

[1] If tropine had one asymmetric carbon, atropine could be depicted as (+ +), the signs representing the configurations of the two asymmetric centers, one in the tropic acid moiety, the other in the tropine. Equilibration of the tropic acid part would lead, not to racemic atropine, but to atropine (+ +) and "iso-atropine" (+ −), both optically active. If the tropine center happened to equilibrate under the same conditions, then *two* optically inactive compounds could result, i.e., *racemic* atropine (+ + and − −) and *racemic* "iso-atropine" (+ − and − +).

Tropine was known to be saturated and contain a tertiary N—CH$_3$ (methylation yields a quaternary salt containing only one more carbon and two Hofmann degradations yield (CH$_3$)$_3$N) and a secondary hydroxyl since it could be reversibly oxidized to a ketone, tropinone. The hydroxyl is the site of tropic acid esterification in the natural alkaloids. The keto group in tropinone is flanked by two α-methylenes as witnessed by formation of a bisbenzylidene derivative. (■)

These facts require that tropine contain two rings and a plane of symmetry; since the alcohol and nitrogen are each only one of a kind, they must lie *on* the plane of symmetry so that the partial formula [11] must be inferred. From this formula only a very few structures can be constructed. A number of other observations are fitted only by the structure [12] for tropine (cf., zinc dust distillation of N-desmethyl-tropane yields 2-ethyl-pyridine), and this was vindicated by Will-stätter's first synthesis, which laboriously carried cycloheptanone into tropine in eighteen steps. The indicated stereochemistry of the hydroxyl shown in [12] follows from the higher pK$_a$ of tropine than the epimeric ψ-tropine.

In the first laboratory synthesis consciously modeled after *in vivo* biosynthesis, Robinson in 1917 mixed succinic dialdehyde, methyl-amine, and acetone in water for thirty minutes at room temperature and isolated tropinone. The spectacular success of this synthesis compared to that of Willstätter vividly argued the value of biogenetic schemes and spawned much subsequent biogenetic speculation. If the acetone be replaced by salts of acetonedicarboxylic acid, remarkably good yields (90 per cent) of tropinone can be obtained. Replacement of succinic with glutaric dialdehyde similarly affords the alkaloid pseudo-pelletierine [2–112].

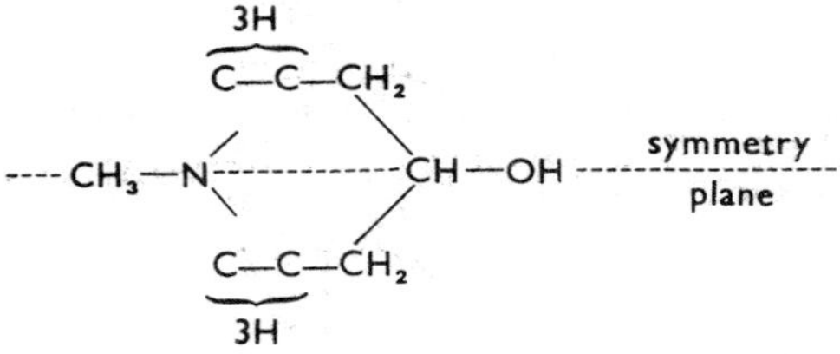

[11]

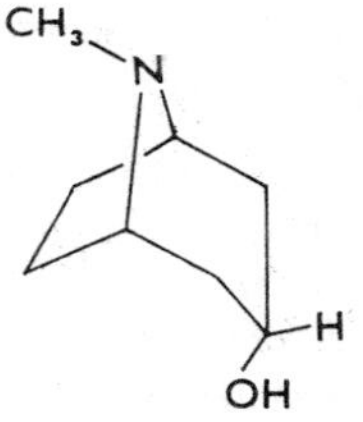

[12]

Tropine

5–2   Phenylalanine Alkaloids

A. GENERAL   Structures for a number of the simpler alkaloids derived from two phenylalanine units were deduced in the later years of the last century and afforded the first hints of modern biogenetic understanding. The degradative methods used were simple in essence and leaned heavily on the synthesis of degradation products. As these compounds always contained at least two aromatic rings, the most rewarding reaction was probably hot $KMnO_4$ oxidation. This is vividly illustrated by the mixture of products, [13]–[15], which was obtained from papaverine ($C_{20}H_{21}NO_4$) and serves in itself to define

the molecule completely (■) once the structures of the several acids have been ascertained by synthesis. Papaverine [16] was the first of the opium alkaloids to yield its correct structure (1888).

The Hofmann degradation has also been used to advantage and always affords first the stilbenes from benzylisoquinolines or the phenanthrenes from aporphine alkaloids (see Chapter 2—Chart 10b). Subsequent oxidation at the generated olefin as well as a second Hofmann procedure to eliminate nitrogen and provide a readily oxidized vinyl group endow this widely used degradative approach with great potency in this family.

In view of the common occurrence of phenols among the alkaloids of this family, the third widely useful degradative device was that of ethylation both to protect and to mark the phenolic group(s). The ethoxyl positions in the degradation products may then be distinguished from the methoxyls (the only alkoxyl groups in the natural molecules) by comparison with synthetic samples and hence the original phenolic sites in the alkaloid discerned. Thus, while the structure of laudanosine [17] is known from reducing and methylating papaverine, structures of the several phenolic alkaloids (cf. laudanine [17]) which yield laudanosine on *methylation* are determined via *ethylation* followed by degradation. With few exceptions these three degradative tools and subsequent synthesis served for the inference and proof of structure of all the many alkaloids derived from two phenylalanine units, except the morphine family.

[16]
Papaverine

[17]
Laudanosine, R = CH₃
Laudanine, R = H

B. MORPHINE   Morphine was the first alkaloid to be recognized as an "alkali" (1803). Its analgesic and euphoric properties have always made it the most-used alkaloid in commerce, and its remarkable and fascinating reactions have made it the most studied alkaloid in chemistry, the subject of over 500 papers in a century and a half. Progress on the deduction of the constitution of morphine, while it contributed much to chemistry generally, did not run a straight or a short course, and the structure proof outlined here is a stringent, *ex post facto* selection of observations which allow the characterization.

It was known early that morphine ($C_{17}H_{19}NO_3$) is a tertiary amine which has two –OH groups (cf. formation of a basic dibenzoyl derivative), one of which is phenolic, the other an aliphatic secondary alcohol oxidizable to a ketone. The alkaloid codeine was related as the phenolic methyl ether of morphine, and the corresponding ketone codeinone is the product of acid hydrolysis of thebaine, its enol methyl ether. Thus these three morphine alkaloids were early interrelated for use in structure deductions.

Hofmann degradation of codeine yielded a basic methine, from which the phenanthrene [18] could be obtained (along with ethylene gas and trimethylamine) after a second Hofmann procedure. The original methine, however, also yielded β-dimethylamino-ethyl acetate and another phenanthrene ([19], R = H), when treated with acetic anhydride, while the corresponding first methine from thebaine yielded ([19], R = OCH₃) instead, These products contain all the carbons of the alkaloid, and Pschorr developed his excellent phenanthrene synthesis at the turn of the century to confirm their structures.

Finally, codeine possesses a double bond, and its allylic relation to the secondary alcohol grouping was shown (*inter alia*) by acid equilibration to four codeine isomers, two of which are oxidized to codeinone, the other two to the isomeric pseudocodeinone. This requires the presence of the grouping –CHOH—C=CH–.

These observations, taken together, create the part structure [20] for morphine and focus attention on the mode of attachment of the ethanamine chain which is so easily eliminated. Many bonding sites for this chain had been proposed, but it was Robinson who recognized as early as 1923 that loss of the fragment was always accompanied by aromatization of the upper ring. He reasoned that the cleavage of the ethanamine chain was concerted with, and gained driving force from,

[18]   [19]   [20]

the aromatization, and concluded that the chain must be attached at a quaternary site, selecting C-13 partly on biogenetic grounds.

The point of attachment of the nitrogen was deduced from CrO_3 oxidation of codeine to a hydroxy-codeine—the hydroxyl of which was at or adjacent to the point of attachment of nitrogen since the Hofmann degradation yielded a ketone—transformed by subsequent reaction with acetic anhydride to a diacetoxyphenanthrene analogous to ([19], R = H), and yielding the *same* phenanthroquinone. Thus the nitrogen is linked to C-9 or C-10 and C-9 was chosen by analogy with the structure of apomorphine [28] (R = H, see p. 153) (and, biogenetically, with laudanosine) and by virtue of its lack of benzylamine properties. Several syntheses have confirmed the formulas [21] for morphine and codeine and [22] for thebaine so derived by Robinson in 1925.

The mystery surrounding the products of Grignard reagents on thebaine, however, delayed general acceptance of the formula for

another twenty years until Robinson, brilliantly interpreting the extensive experimental work of Small and others, uncovered the remarkable
explanation. Phenyl magnesium bromide converts thebaine into two
isomeric " phenyldihydrothebaines " which are diastereomers but each
is transformed into the *enantiomer* of the *other* on heating! The
phenyldihydrothebaine diastereomers are tertiary amines but are both
converted to the *same* hydrogenolysis product (one mole/H_2), a
secondary amine. Hydrogenolysis of the two pyrolysis enantiomers
yields the enantiomer of this secondary amine. Similarly, two successive Hofmann degradations afford the same nitrogen-free but still
optically active diolefin from each isomer. Especially puzzling was the
evident asymmetry in these compounds despite much evidence that

[21]
Morphine, R = H
Codeine,   R = CH_3

[22]
Thebaine

the upper ring had gone aromatic in the Grignard reaction. Robinson's
solution is found in the following paragraph. (■)

Much of the confusion in deducing the attachment of the ethanamine chain had arisen from the number of rearranged products in
which that chain was linked to other sites. The rearrangements of
thebaine are all initiated by attack of an electron-deficient species (as
$H^{\oplus}$) on the oxide to form the allylic carbonium ion [23] which
rearranges to the more stable [24]. The latter ion has no protons it
can now lose but it may accept external electrons, as in the $SnCl_2/HCl$
reduction to metathebainone [26], or may collapse to the more stable
cyclic immonium salt [27] by incursion of the unshared electron pair
from nitrogen. This functionality now affords an active site for
attack of the Grignard carbanion to yield the alkyl- or phenyldihydrothebaines [29]. The two diastereomers formed represent the
two possible configurations at the asterisked site, superimposed on
the *molecular* asymmetry of the biphenyl system in which free rotation
is severely hindered.

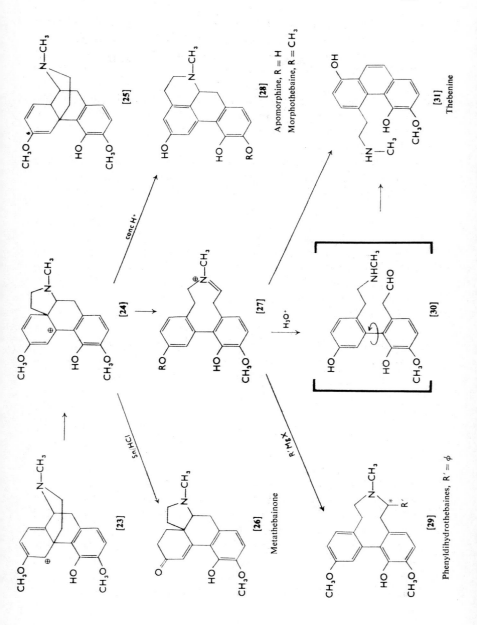

[25]

[28]
Apomorphine, R = H
Morphothebaine, R = CH₃

[31]
Thebenine

conc H⁺

[24] →

[27]

H₃O⁺

[30]

[23]

Sn/HCl

[26]
Metathebainone

R'MgX

[29]
Phenyldihydrothebaines, R' = φ

Heating the isomers [29] results in "epimerizing" this molecular asymmetry to the opposite configuration. If the biphenyl configuration is denoted by ± and the asterisked asymmetric carbon configuration by R or S, the two diastereomers are describable as (R, +) and (S, +) and yield, respectively, (R, −) and (S, −), i.e., their mutual enantiomers, on heating. Double Hofmann or hydrogenolysis procedures destroy the asymmetric carbon but still leave the optical activity of the hindered biphenyl with three ortho substituents.

In dilute aqueous acid the immonium salt ([27], R = H) is hydrolyzed to a transient aldehyde [30] which is free to rotate and condense (ortho to the phenolic –OH), forming the phenanthrene thebenine [31]. In strong acid, however, the intermediate [24] is largely protonated on nitrogen and so must rearrange further in order to achieve a species which can collapse by deprotonation to yield morphothebaine ([28], R = CH₃). The structure of apomorphine ([28], R = H), the corresponding derivative from morphine and strong HCl, was determined and synthesized long before that of morphine itself and influenced deductions about the latter as well as providing a model for elucidating the natural aporphine alkaloids.

The stereochemistry of the five asymmetric centers in morphine was brilliantly deduced by Stork, who applied modern stereoelectronic considerations to previously available data. Heating codeine methyl ether with sodium ethoxide causes elimination of the phenoxide ion and affords [25], but the epimeric (at C-6) methyl ether is unreactive. Hence the C_6—H is *trans* to the oxide bridge in codeine (and morphine) [21]. The five membered oxide ring is then fused either *cis* or *trans* to the top, hydroaromatic ring, and its refusal to epimerize at C-5 via enol equilibration of dihydrocodeinone implies that codeine possesses the more stable *cis* ring junction.

Hence the ethanamine chain and the oxide link are oriented *trans*, consistent with those reactions in which both are eliminated together, as in the acetic anhydride treatment of the Hofmann methine product [32] from codeine. The ethanamine bridge across the six-membered ring can, furthermore, only be *cis* at C-9. The remaining center at C-14 is fixed by hydrogenation of thebaine, which affords dihydrocodeine methyl ether quantitatively, placing *cis* hydrogens at C-6 and C-14. The resultant steric projection is shown as [33] and is consistent with the X-ray crystallographic determination; the absolute configuration is that shown in [33].

The relation of sinomenine [34] to morphine is interesting, for Goto

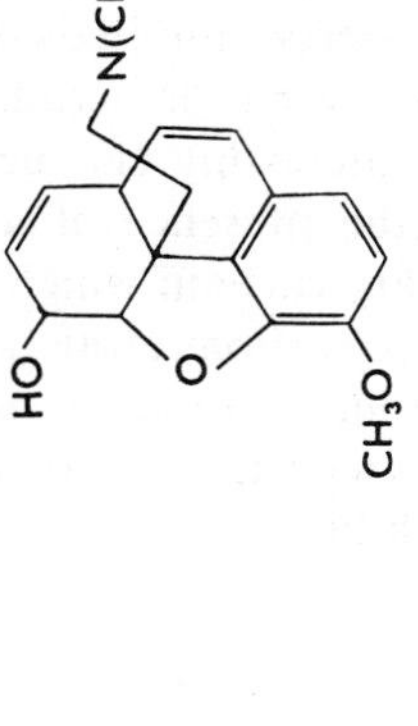

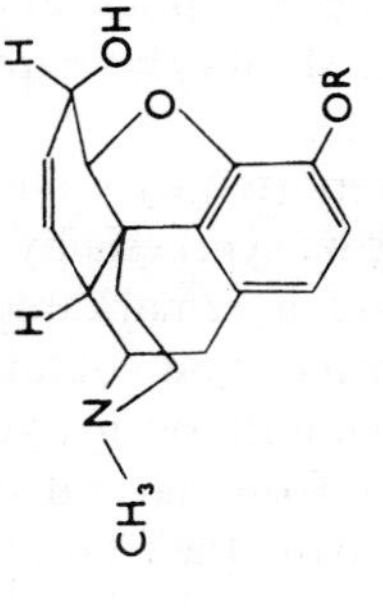

[33]

Morphine, R = H
Codeine, R = CH₃

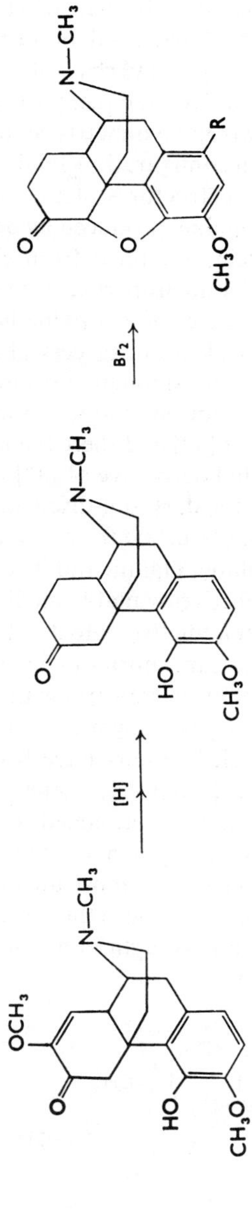

[36]

[35]

[32]

[34]
Sinomenine

showed that reduction of sinomenine to [35], followed by bromination, afforded ([36], R = Br), which in turn yielded a hydrogenation product identical with dihydrocodeinone ([36], R = H) in every respect except that it had an equal and opposite optical rotation. Mixture of these two natural derivatives in equal parts leads to racemic dihydrocodeinone. Therefore, *every* asymmetric center in this product and in sinomenine is inverted in configuration relative to its counterpart in the otherwise identical structures of the morphine family. The enantiomer of dihydromorphine, produced by demethylating the optical antipode of dihydrocodeine obtained from sinomenine, shows the narcotic but not the analgesic properties of morphine.

Total syntheses of morphine have been reported by Gates and by Ginsburg, while an unsuccessful attempt at its synthesis by an oxidation meant to simulate its biogenesis had a remarkable sequel. Robinson oxidized N-methyl-norlaudanosine and obtained not morphine but [37] and there the matter rested until twenty years later a tri-O-methyl derivative of [37] was isolated from natural sources!

Opium is the dark sun-dried latex exuded from the fruit of *Papaver somniferum*, about a fifth of which consists of alkaloids, half of this being morphine. Opium and its constituents have always been by far the most effective available medicines for the relief of pain, but their concomitant addictive potential has always loomed as a serious drawback, so that an enormous chemical effort has been expended to create morphine substitutes without this property, hundreds of compounds having been synthesized and tested over four decades. The pharmacological results have led to some interesting chemical correlations, viz., analgesic activity requires the presence of a *meta*-hydroxy-phenyl ring attached to a quaternary carbon, which is part of a ring containing an alkylated nitrogen two carbons away from the quaternary center. Almost all these compounds, however, are also narcotic. An exception is the synthetic [38], which appears to be (or to lead to) the long-sought nonaddictive pain-killer.

[37]

[38]

C. AMARYLLIS ALKALOIDS   The poisonous bulbs of Narcissus and Amaryllis plants were first examined for alkaloids in the nineteenth century but inspired only sporadic study until the 1950's, when a number of groups, most notably that of Wildman, carried out the

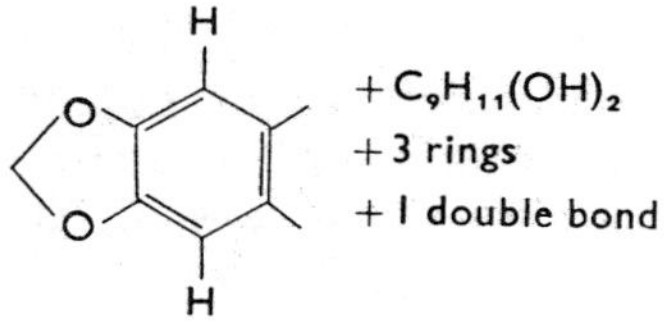

$+ C_9H_{11}(OH)_2$
$+ 3$ rings
$+ 1$ double bond

[39]

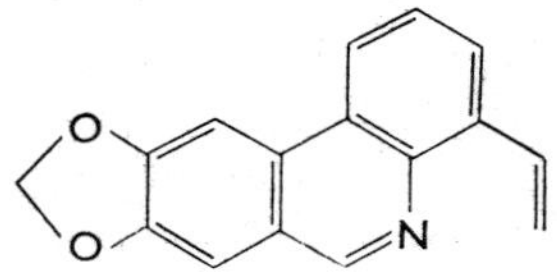

[40]

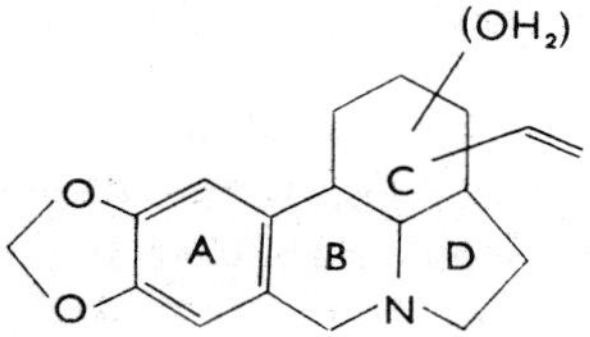

(OH_2)

[41]

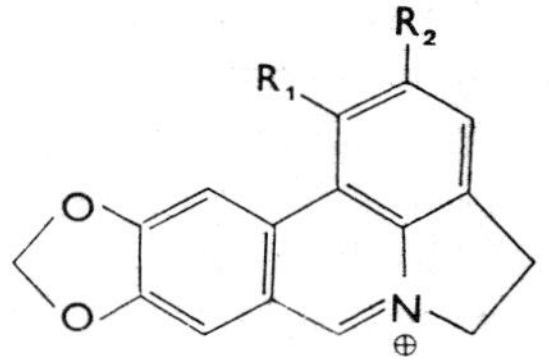

R_2
R_1

[42]

a: $R_1 = H$; $R_2 = O^{\ominus}$
b: $R_1 = O^{\ominus}$; $R_2 = H$

HO
R

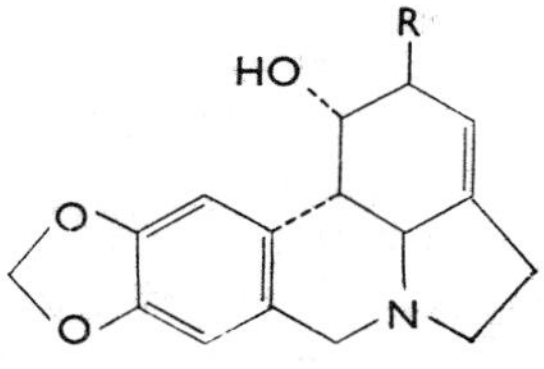

[43]

Lycorine: R = OH
Caranine: R = H

extensive examinations that have resulted in a rich collection of over seventy-five alkaloids and their structures. This rapid achievement owes much of its momentum to the possibility of the conversion of new alkaloids into known degradation products of the parent—the first member of the family to be elucidated—as well as to the potency of UV and IR spectra for locating functional groups. The two major structural families (see Chapter 2) are represented by their parent alkaloids, lycorine and crinine.

Kondo developed most of the structure proof for lycorine in the 1930's by classical means. The formation of diacetyl and dihydro derivatives, the presence of a tertiary amine (not N—CH₃), the absence of phenolic or ketonic reactivity, and finally the production of 4,5-methylenedioxy-phthalic acid from hot $KMnO_4$ oxidation accounted for all the functionality implied in the composition, $C_{16}H_{17}NO_4$, of lycorine. It must therefore have three nonaromatic rings in addition to the methylenedioxyphenyl system [39]. Hofmann degradation leads to a vinyl compound which has also lost the alcohol groups and is optically inactive, suggesting aromatization of another ring.

The Hofmann product is in fact [40], as Kondo deduced by removal of the vinyl group and synthesis of the resultant aromatic, as well as by hydrogenation and zinc-dust distillation to the 1-ethyl-phenanthridine. The formula for lycorine may thus be expanded to [41] and placement of the two –OH groups and the olefin in ring C afford it the same oxidation state as an aromatic ring, which accounts for its ready dehydration and aromatization in the Hofmann and a variety of other reactions.

On mild oxidations lycorine yields both internal salts [42], which are of great value for establishing hydroxyl positions in members of the lycorine family because of their different and characteristic UV spectra. Taking the hydroxyls now at the substituted positions of ring C in [42], only the double bond must be placed in that ring to complete the lycorine structure. The identity of the UV spectrum of lycorine with that of simple methylenedioxphenyl compounds precludes the two conjugated positions and the greater basicity of dihydrolycorine (pK$_a$ 8·4) over lycorine (pK$_a$ 6·4) is not consonant with an eneamine (N—C=C) system.

Hence, as lycorine is not an enol, the only place remaining in ring C for the double bond is that shown in [43]. This has been further confirmed by a number of other reactions including $Pb(OAc)_4$ cleavage of the α-glycol and MnO_2 oxidation of the allylic alcohol to ketone. The *trans*-diaxial nature of the glycol is shown by the facile production of an epoxide when a monotosylate of dihydrolycorine is treated with base, the glycol being then regenerated by $H_3O^\oplus$. This information is implicit in the full stereochemistry shown in [43].

In contrast to lycorine, crinine ($C_{16}H_{17}NO_3$) was found to be stable to many oxidative and dehydrogenative reagents and the Hofmann degradation proceeded poorly. Besides the methylenedioxyphenyl system and tertiary amine, crinine had, like lycorine, an allylic alcohol, attested by MnO_2 oxidation to an isolated unsaturated ketone system with its characteristic UV absorption (226 mμ). Its co-occurrence in the plant with lycorine and its close similarity in empirical formula and functionality implied that they were biogenetically closely related,

[44]

while the resistance to aromatization suggested a quaternary carbon in ring C. Wildman accordingly postulated the structure [44] for crinane, the parent skeleton obtained by reductive removal of the allylic alcohol (MnO_2; H_2/Pd; $H_2NNH_2/OH^{\ominus}$), and confirmed it by synthesis. The position of the allylic alcohol was elegantly demonstrated by MnO_2 oxidation to oxocrinine, and its facile Hofmann degradation to an optically inactive dienone. (∎) Thus, crinine must be [45].

Once these parent structures were established the interconversions of their degradation products and newly isolated alkaloids usually served to identify the latter. Certain reactions had widespread utility for these interconversions: (a) MnO_2 oxidation of allylic alcohols to unsaturated ketones, identified by UV spectra; (b) solvolytic interconversion of allylic alcohol and corresponding methyl ether; (c) mild oxidations in the lycorine family to betaines (cf. [42]), identifying positions of oxygen attachment by UV absorption; (d) removal of aromatic or allylic methoxyl groups by Na/ROH reduction.

Thus caranine is similar to lycorine with one less oxygen and yields [42b] on oxidation. Its identity with ([43], R = H) and synthesis from the lycorine epoxide (above) with $LiAlH_4$ followed directly. The demethoxylations are represented by the Na/ROH reduction of falcatine [46] to caranine and the conversion of galanthine [47] to pluviine [47]. A much more spectacular transformation is that of haemanthidine [2–138] to tazettine [48] by methyl iodide; an internal Cannizzaro-type hydride transfer is involved.

5–3   Indole Alkaloids

A. CALYCANTHINE   Calycanthine, $C_{22}H_{26}N_4$, was the first of the *Calycanthus* alkaloids to be isolated; chemical degradations in the 1930's attested to the complexity of its structure but added little to its

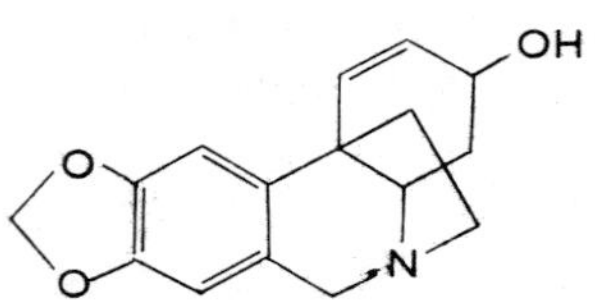

[45]
Crinine

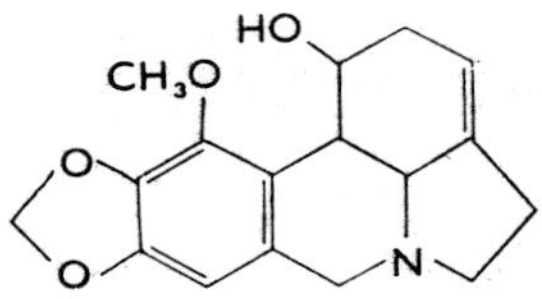

[46]
Falcatine

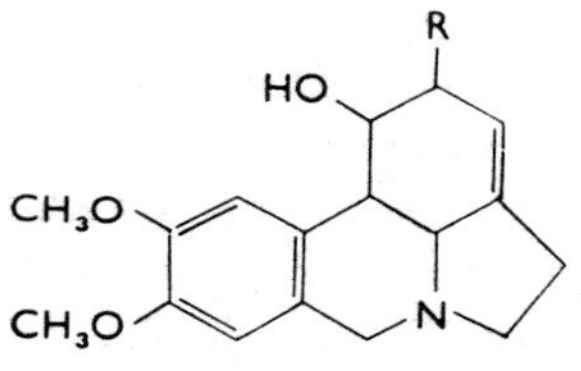

[47]
Galanthine, R = OCH₃
Pluviine, R = H

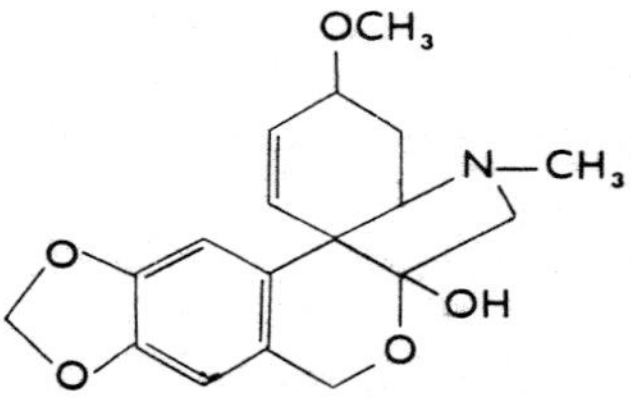

[48]
Tazettine

unveiling. The solution of the problem nicely exemplifies the potency of applying both mechanistic and biogenetic theory to structural inference. The isolation of N-methyltryptamine in good yield from a number of vigorous degradations, and the presence of calcyanthine in two taxonomically well-separated species led Woodward to postulate a simple biogenesis from N-methyltryptamine. Furthermore, the empirical formula of calcyanthine is just twice that of methyltryptamine minus two hydrogens and so suggested an origin in a simple oxidative dimerization.

On mechanistic grounds the primary availability of the indole-nitrogen electrons at the β-coupled oxidative dimer [49] of N-methyltryptamine as the biogenetic precursor. The dimer [49] is an isomer of calcyanthine and spontaneous cyclization of the aliphatic amino groups to the imines can produce two hexacyclic formulas, one of which [50] has since been found to be the structure of chimonanthine, a minor alkaloid from the same plant. Furthermore, simple hydrolysis of the two imine groups in [49] leads to a dialdehyde tetraamine [51] which may now reclose to yield only five possible saturated isomers of $C_{22}H_{26}N_4$, one of which is chimonanthine, another, calcyanthine [52].

A very simple synthesis, modeled after the biogenesis, consists of FeCl₃ oxidation of the indolic anion (as the Grignard) of N-methyl-tryptamine to chimonanthine in 20 per cent yield, whereas acid hydrolysis of either chimonanthine or calycanthine leads to an equilibrium mixture of the two, presumably through the intermediacy of the aldehyde [51] and similar species. The fact that calycanthine is optically active allows the deduction of its stereochemistry directly.

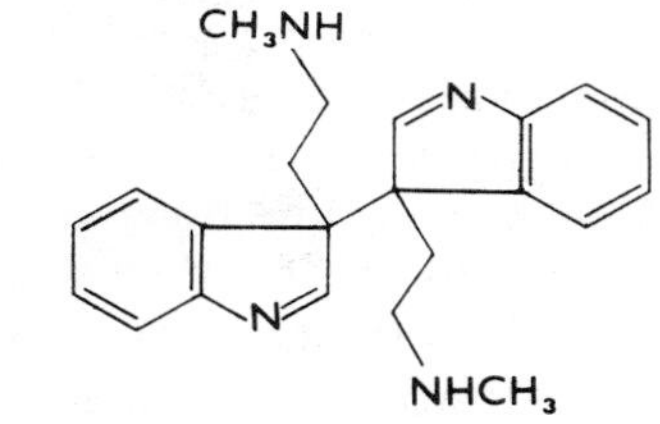

[49]

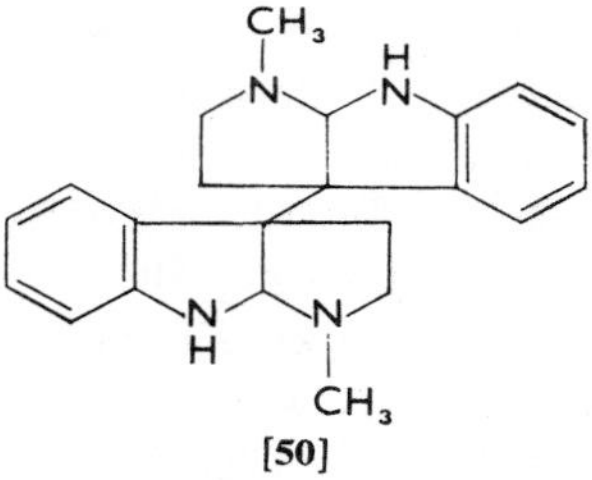

[50]
Chimonanthine

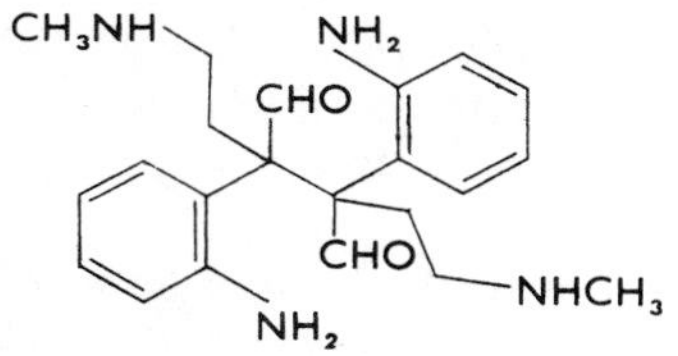

[51]

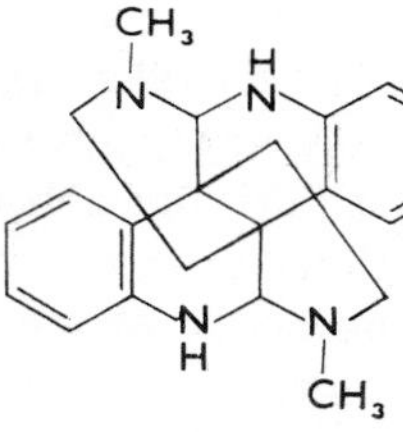

[52]
Calycanthine

B. QUININE   Legend has it that the Countess Chinchon, wife of the viceroy of Peru, was the first European cured of malaria by the bark of the *quina-quina* tree in 1638 and subsequently introduced it into medical practice. In any case, Linnaeus bestowed her (misspelled) name on the *Cinchona* species, and Pelletier and Caventou isolated quinine from its bark in 1820. The medical importance of the drug and its availability in quantity (the Dutch cultivated *Cinchona* trees in Java and bred alkaloid concentrations as high as 20 per cent in the bark) led to very active chemical study in the nineteenth century, culminating not only in the correct structure by 1907, the earliest of the complex alkaloids, but also in a vast clarification of pyridine and quinoline chemistry.

The analysis of Pelletier in 1823 was correct but it was 1854 before sufficient comprehension of valency allowed Strecker to convert this to the correct molecular composition, $C_{20}H_{24}N_2O_2$. At almost the same time Pasteur carried out some of the first important degradations of quinine and also used it in the first application of his celebrated resolution procedure to separate the enantiomers of DL-tartaric acid. The chemistry of quinoline began with its production from KOH fusion of the alkaloid cinchonine by Gerhardt in 1842 and its development was closely bound to the problem of the structure of the *Cinchona* alkaloids. Finally it may be said that the commercial production of synthetic dyes, the first and for a long time the major chemical industry, sprang from Perkin's famous if naïve attempt to synthesize quinine in 1856, when he obtained instead the dye "mauve" which soon became the rage of *fin-de-siècle* fashion.

Strecker early showed that quinine and cinchonine, the two major alkaloids of *Cinchona*, had two tertiary nitrogens by their transformations into two different monoethiodides. Since acetic anhydride yielded a basic monoacetate of cinchonine, its one oxygen was a hydroxyl (not phenolic by $FeCl_3$). Quinine differs only in having an added methoxyl group, and both have a single double bond, shown to be a vinyl group by oxidation with cold $KMnO_4$ to HCOOH, and to an acid empirically CH_2 less than the parent alkaloid. In this way all the functional groups were defined.

Knowledge of the skeleton began slowly with the formation of quinoline and 4-methyl-quinoline from cinchonine (and their 6-methoxy derivatives from quinine) in various pyrolytic reactions, but little more could be done until the chemistry of these bases was well established and the relevant isomers confirmed by synthesis. In an important sequence studied by Koenigs, quinine and cinchonine were each dehydrated to corresponding olefins which in turn yielded, on mild hydrolysis, two fragments containing all the initial atoms. One fragment was 4-methylquinoline (or 6-methoxy-4-methylquinoline from quinine) which implied the attachment of the rest of the alkaloid at the 4-methyl position, and the other was dubbed meroquinene and was formed from both alkaloids.

Meroquinene, $C_9H_{15}NO_2$, is clearly not aromatic from its composition but contains a carboxyl and secondary amine. Its skeleton is seen in the decarboxylated disproportion product, 3-ethyl-4-methyl-pyridine, formed on heating with HCl in a sealed tube. Meroquinene also contains the vinyl group of quinine since cold $KMnO_4$ transforms it to a diacid with loss of CH_2. Loss of CH_2 again occurs on vigorous oxidation of this diacid to another diacid $(C_7H_{11}NO_4)$; all three acids are also secondary amines. (■) The second oxidation implies $-CH_2COOH \rightarrow -COOH$ so that meroquinene must be [53]. Since meroquinene is formed by hydrolysis (by 2 moles of H_2O), the carboxyl must be the site of its detachment from the 4-methyl-quinoline, and granting the methyl as the quinoline site of attachment, we may attach the two fragments as in the formulas [54] for quinine and cinchonine. Rabe's demonstration (1907) that the hydroxyl could be reversibly oxidized to a ketone allowed him to write these correct structures for the alkaloids. (■) Quinine was synthesized much later, by Woodward in 1944.

The major interesting reactions of quinine, such as Pasteur's conversion to quinotoxine [55] by mild base or the aqueous acid-catalyzed transformation of the dehydrated quinine to meroquinene and apoquinene [56], may generally be formulated mechanistically by regarding the pyridine ring of the quinoline as the nitrogen analog of an α,β-unsaturated ketone.

Of the four asymmetric centers the two which remain in meroquinene are identical in all the natural bases since the same meroquinene is formed from all. The *cis*-fusion of the two side chains on the piperidine ring was shown by Prelog through conversion of meroquinene to optically inactive, hence *cis*, 1,2-diethylcyclohexane. None of the reactions used can epimerize these centers. On the other hand, all four epimers of the other two centers are naturally occurring; and all four can be formed by refluxing quinine in strong base since a trace of ketone allows Oppenauer-oxidative equilibrium of the secondary alcohol with its corresponding ketone in small amount, and the latter provides for equilibrium of the adjacent asymmetric center through its enol.

Two of these epimeric alcohols convert smoothly to cyclic ethers [57], thus establishing their configuration at C-8, the linkage with the azabicyclooctane (quinuclidine) moiety. The two that do not form ethers differ in basicity; as in a number of epimeric pairs of alkaloids with hydroxyls (cf. tropines, p.148), the difference in basicity apparently arises from the difference in ease of hydrogen-bonding of the acidic $N^{\oplus}$—H with the neighboring hydroxyl. This can be seen in the formula and projection down the C_{9-8} bond [58] of epiquinine, the C-9 epimer of quinine. Epiquinine forms the more favorable (less hindered) hydrogen bond; it stabilizes the conjugate acid and so makes epiquinine a stronger base than quinine. The C-9 configurations of the two epimers which do form cyclic ethers were then obtained by comparison of optical rotation differences among the quinine and cinchonine isomers with those among the known isomers in the similar natural ephedrine series, ϕ—CHOH—CH(NHCH$_3$)—CH$_3$.

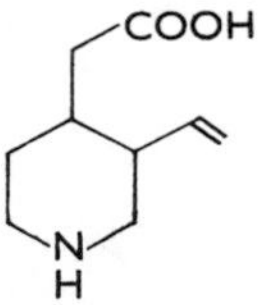

[53]

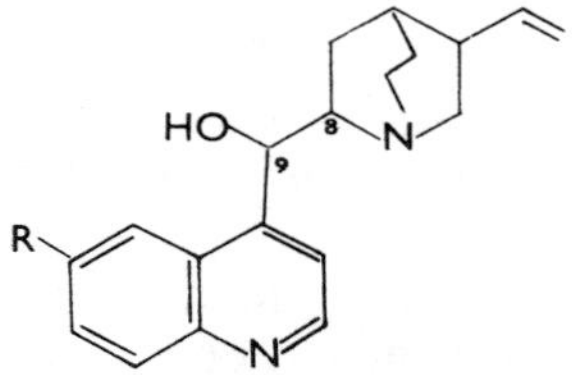

[54]
Quinine.   R = OCH$_3$
Cinchonine,   R = H

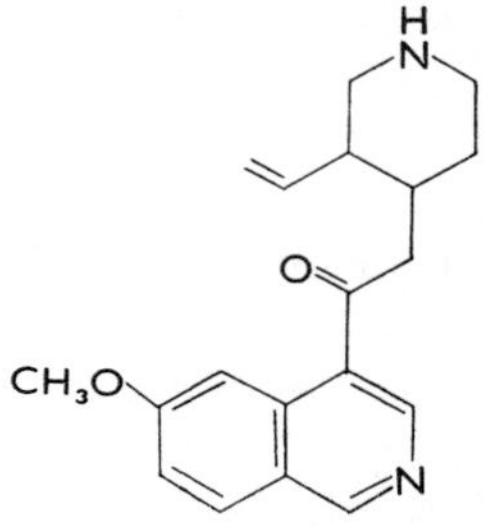

[55]

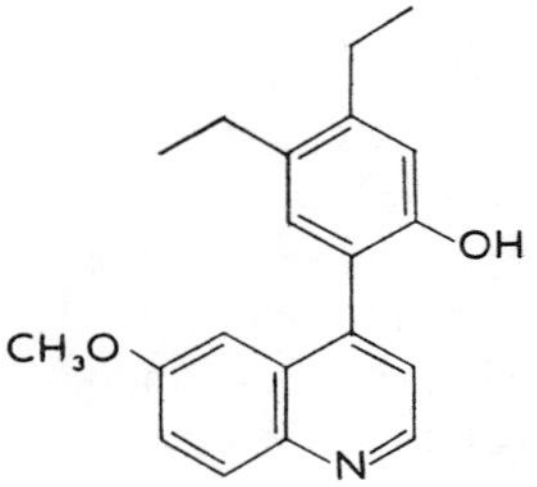

[56]

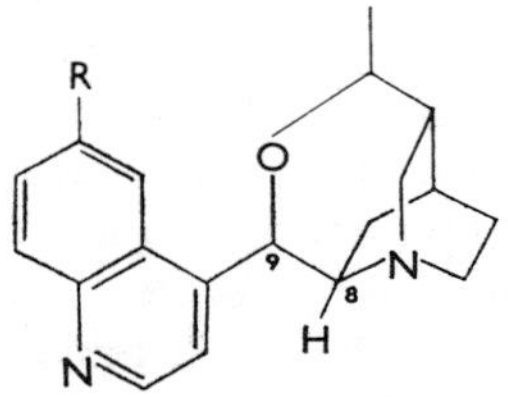

[57]

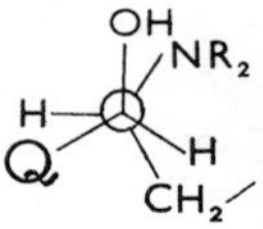

[58]

C. VINDOLINE   A dramatic example of the efficacy of the modern apparatus for structural inference in physical methods, biosynthetic reasoning, and mechanistic understanding, may be found in the recent structure proof by Gorman and Neuss of vindoline, the major one of about sixty alkaloidal constituents so far isolated by careful chromatography from periwinkles (*Vinca rosea*) in a search for anticancer agents. The major contrast with classical structure proofs of comparable molecular complexity lies in the fact that a total of only six chemical degradation reactions were carried out; this may be compared with well over five hundred recorded degradation products of strychnine prior to its ultimate elucidation. It must be emphasized that a most important feature of the modern approach is the possibility of doing a full elucidation with very small amounts of material available.

The empirical formula of $C_{25}H_{32}N_2O_6$ was reinforced by an exact molecular weight from a mass spectrum. The UV spectrum (and its shift in acid) was that of 6-methoxy-dihydroindole and the IR spectrum indicated hydroxyl and ester functions; the latter included one acetoxyl and one carbomethoxy group as shown by $LiAlH_4$ as well as acetylation and deacetylation reactions. The NMR spectrum indicated an olefin which disappeared on catalytic uptake of one mole of hydrogen while an identical UV spectrum in the product, dihydrovindoline, showed the double bond to be a nonconjugated one.

In this way the functional groups ($-N_2O_6$) are accounted for, and an examination of the empirical formula compared to the saturated composition $C_{25}H_{54}N_2$ shows eleven sites of unsaturation, six of which are present in the two ester groupings, and one double bond and benzene ring unsaturation, so that the molecule is pentacyclic. Soda-lime pyrolysis of vindoline yields [59], allowing placement of N—CH$_3$ in the alkaloid and this, coupled with the result that labeled

[59]   [60]

$+ C_3H_3 + 3$ rings more

tryptophan is a good precursor for the alkaloid *in vivo*, affords good evidence for a β-ethanamine side chain on the dihydroindole nucleus.

The NMR spectrum (Table 1) accounts for and distinguishes all thirty-two protons. It reveals an ethyl group attached to carbon (9·52τ, –CH$_3$; 8·65τ, –CH$_2$–) and reveals the acetoxyl methyl (7·93τ), the four protons of the ethanamine side chain (7–7·9τ), the N—CH$_3$ (7·32τ), the two O—CH$_3$ groups (6·20τ), and the aromatic ring bearing three hydrogens as oriented in 6-methoxy-dihydroindole (ortho coupling, $J = 8$ cps; meta coupling, $J = 2·5$ cps). The olefin is seen to be symmetrically substituted with two *cis* ($J = 10$ cps) hydrogens, one of which (4·77τ) is not further coupled and so has no adjacent protons, while the other (4·12τ) is split by the two adjacent but nonequivalent protons whose own absorption occurs at 6·6τ, a chemical shift which requires they be also adjacent to nitrogen. Furthermore, the chain N—CH$_2$—CH=CH—C so deduced can only be placed at the aliphatic nitrogen allowing expansion of the formula to [60], in which x implies a group which cannot be hydrogen.

TABLE I ■

NMR Spectrum of Vindoline

τ	No./H	Splitting	J (cps)	τ	No./H	Splitting	J (cps)
9·52	3	3	7·5	6·20	6	1	...
8·65	2	M*	7·5	4·77	1	2	10
7·93	3	1	...	4·57	1	1	...
7·0–				4·12	1	2×2×2	2, 5, 10
7·9	4	M*	...				
7·35	1	1	...	3·92	1	2	2·5
7·32	3	1	...	3·70	1	2×2	2·5, 8
6·6	2	2, 2	2, 5	3·09	1	2	8
6·25	1	1	...	1·00	1	1	...

* M = multiplet.

Several conclusions about [60] are possible: (1) the three remaining protons seen in the NMR are all unsplit and at rather low field, hence none are at the β-position of the dihydroindole (marked with x); (2) the proton at 7·35τ is an unsplit –N—C—H and so must be adjacent to one of the two nitrogens; (3) the three functional groups must be close together since there remain only three skeletal carbons to place.

Dihydrovindoline hydrochloride on mild pyrolysis affords a fair yield of a ketone, C$_{21}$H$_{28}$N$_2$O$_2$ (IR absorption at 5·85 μ indicates a cyclohexanone), with the UV spectrum of vindoline. The ketone carbonyl is flanked by two α-hydrogens as shown by an increase of two in the molecular weight from the mass spectrum after deuterium equilibration of the enol in CH$_3$O$^⊖$/CH$_3$OD. Since analysis of the sites of unsaturation shows a skeleton with the same number of rings (five), we may conclude that heating the hydroxyl, acetoxyl, and carbomethoxy functions of dihydrovindoline with a mole of HCl affords a ketone. On mechanistic grounds this can be accommodated in two ways [61]. In the NMR spectrum the —CH$_2$—CO— group (readily identified by its disappearance on deuteration) exhibits a doublet split by a single proton at 7·5τ (which goes to a singlet on deuteration) and so next to nitrogen, implying N—CH—CH$_2$—CO— in the pyrolysis product. Therefore, in vindoline the corresponding array must be [62], the proton at (a) being assigned 6.25τ, the one at (b) 4.57τ and so too low for –CH—OH.

If one now carries out the very interesting intellectual exercise of putting these pieces together to form three new rings (without changing the neighboring patterns of protons from those implied in their

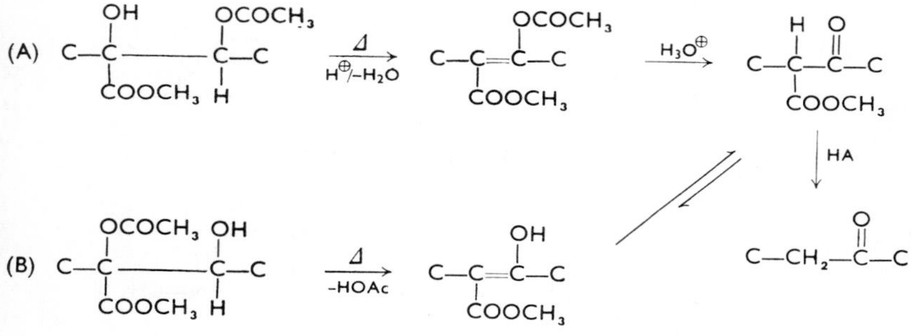

[61]

NMR splitting and placing the pyrolysis ketone in a six-membered ring) (■), then it will be found that only [63] and [64] are viable formulas. The abnormally high field of the methyl protons in the ethyl groups—implying as it does a close proximity to the π-electron current of the aromatic ring—both eliminates [64] and requires a cis-1,3-diaxial orientation of the aromatic ring and the ethyl group

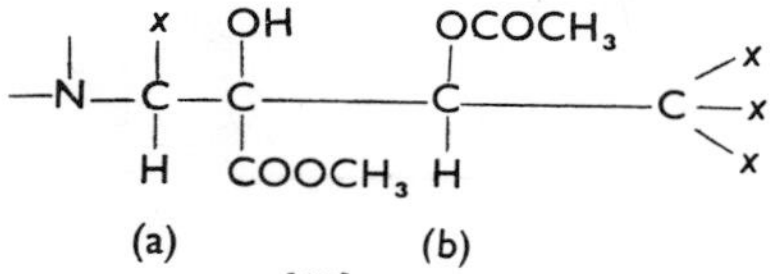

[62]

relative to the cyclohexane ring to which they are both attached in [63]. Mass spectral cracking patterns of the pyrolysis ketone confirmed its skeletal identity with the known alkaloid aspidospermine [2–158], which has been synthesized by Stork.

D. AJMALICINE AND MITRAPHYLLINE The progression of ideas on alkaloid biogenesis has always been marked by laboratory syntheses reflecting these ideas, Robinson's early tropinone preparation (p. 148) having provided the impetus. The syntheses in the 1930's of [65] from tryptamine, formaldehyde, and various phenylacetaldehydes under very mild conditions afforded the skeleton of the then-known indole alkaloids (cf. yohimbine [2–144]) in which the non-tryptamine portion was believed to arise from phenylalanine.

A recent synthesis of ajmalicine [71] by van Tamelen illustrates the application of this approach to newer biogenetic ideas. A simple

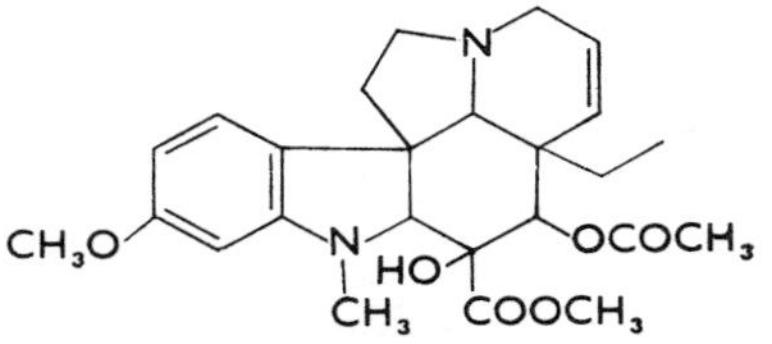

[63]
Vindoline

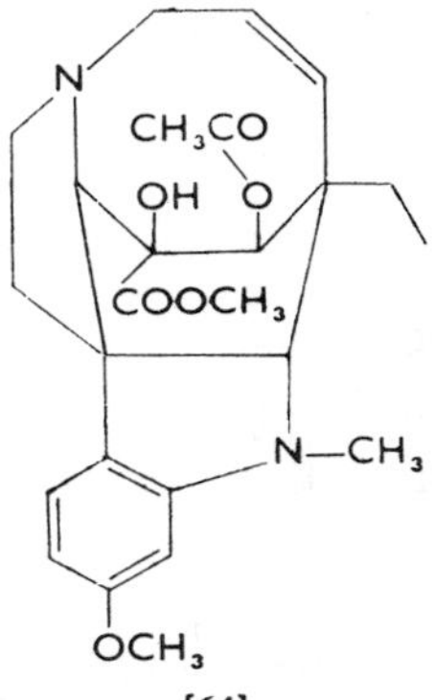

[64]

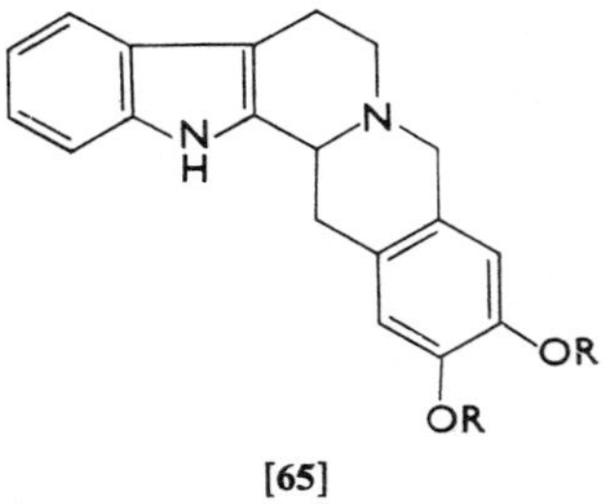

[65]

Mannich condensation of tryptamine, formaldehyde, and [66], formed by Michael addition of acetoacetic ester to glutaconic (pentenedioic) ester, yielded the lactam [67], the ester of the initial Mannich product having cyclized spontaneously. The power of the biogenetic model (cf. [2–145]) for synthesis will be seen in the fact that this molecule [67] now contains virtually the complete skeleton of ajmalicine [71].

After cyclizing this amide in [67] to the indole ring (with POCl$_3$) and

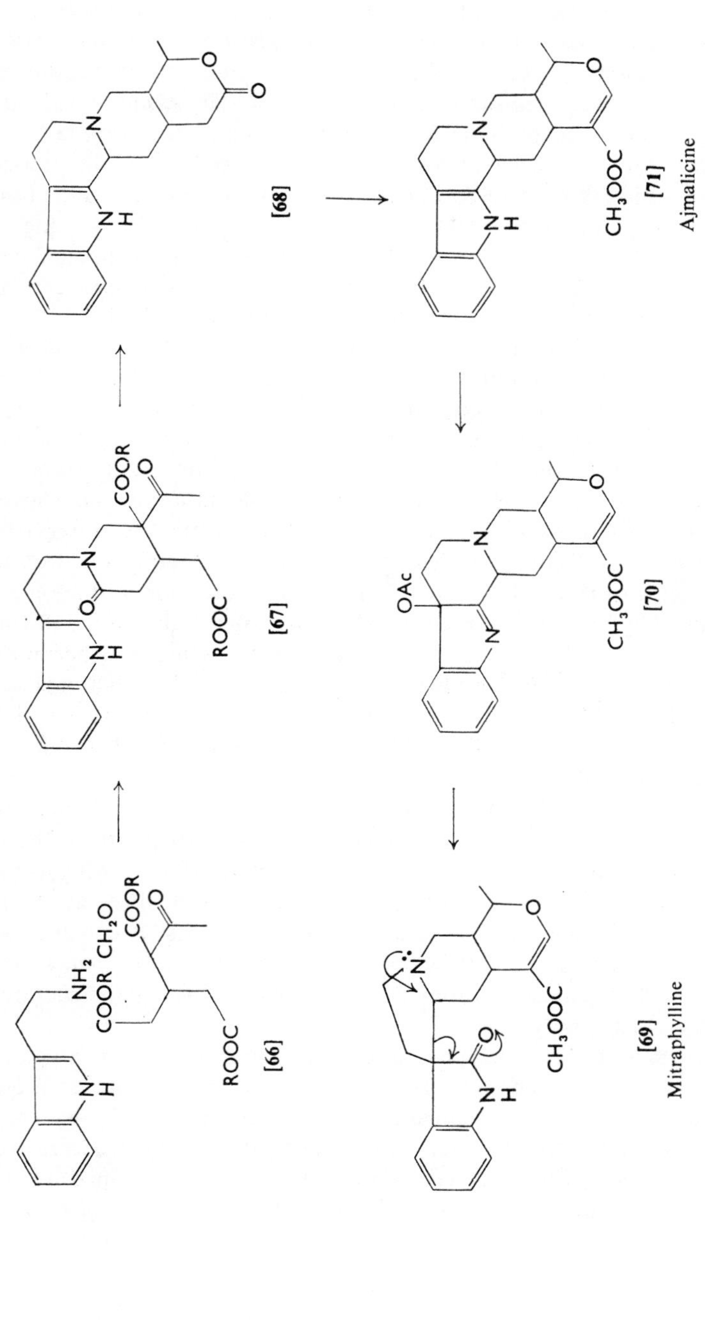

[68]

[71]
Ajmalicine

[67]

[70]

[66]

[69]
Mitraphylline

hydrogenating the immonium salt formed, the product was submitted to acid hydrolysis which causes decarboxylation and epimerization of the asymmetric centers adjacent to indole and ketone, producing the most stable stereoisomer. The subsequent elaboration of the fifth ring was achieved by hydride reduction of the methyl-ketone to an alcohol which cyclized directly to a lactone [68], allowing Claisen condensation (with ethyl formate) to the α-hydroxymethylene-lactone and subsequent lactone opening and closure of an enol ether ring by acidic methanolysis to form ajmalicine. As an amusing footnote, it may be said that the correct stereochemistry at the four asymmetric centers was not known prior to this total synthesis of the alkaloid.

A group of alkaloids containing oxindole rings has been found to be closely related to certain corresponding indole alkaloids, as may be appreciated in the structure of mitraphylline [69], comparable to that of ajmalicine [71]. These oxindole alkaloids differ only in the fifth ring (or in the aromatic substituents) and show an interesting isomerization on heating, apparently caused by the tautomerism shown (arrows in [69]). As this reaction destroys two asymmetric centers (the three associated with the fifth ring are unaffected), their reconstitution in the tautomeric equilibrium can in principle afford four diastereomers of mitraphylline. In fact the two which are most stable on conformational grounds are formed as an equilibrium mixture. Furthermore, one of the central degradation products of these oxindole alkaloids was the cyclopropane pyrolysis product, 3,3-dimethylene-oxindole, which may be seen to be readily derivable from the tautomer of [69] produced by the arrows shown.

Inasmuch as this isomerization of the two centers is available, and natural mitraphylline is one of the two isomers formed, its synthesis from ajmalicine may be considered. Reaction of indoles with various oxidants (cf. $Pb(OAc)_4$) leads to 3-substituted indolenines, as in the product [70] from ajmalicine, and these in turn easily rearrange to oxindoles. Warming [70] with aqueous methanol in fact yields mitraphylline and isomitraphylline and so establishes their structures and stereochemistry.

E. RESERPINE   The conspicuous clinical importance of the alkaloid reserpine as a hypotensive and tranquilizer has in recent years sparked a renaissance in the isolation and study of indole alkaloids which has notably enriched the chemistry of natural products. The structure and stereochemistry [72] of reserpine itself were both determined within about two years of its isolation in crystalline form, and these were followed in only one more year by a brilliant synthesis in Woodward's laboratory in 1955, probably the most elegant synthesis of a complex alkaloid ever achieved (Chart 1). Since the six asymmetric centers in reserpine imply thirty-two possible racemic products from a sterically undirected synthesis, the central synthetic consideration must be complete stereochemical control in every reaction, especially since in the natural alkaloid not all the centers have the thermodynamically most stable orientation.

As the D/E-ring system bears a *cis*-fusion and such fused rings (unlike *trans*-fused) are as conformationally mobile as cyclohexane itself, synthetic stereochemical control may be achieved by locking the rings into only one of their two possible chair–chair conformations. Furthermore, in such *cis*-fused bicyclic systems, the molecule has a very cupped shape so that its convex face (the side of the two *cis*-hydrogens at the ring junction) is very much more accessible to approaching reactants and so allows a special degree of stereoselective control. Accordingly, a *cis*-decalin (decahydronaphthalene) system was taken as a starting point to elaborate the stereochemical features of the D/E-ring sector of reserpine.

The *cis*-decalin [73], created with the stereochemistry shown by the Diels-Alder reaction of benzoquinone with butadiene-1-carboxylic acid ester, was reduced by the aluminum isopropoxide procedure, with approach of hydrogen exclusively from the convex side, yielding a diol which cyclized spontaneously to the lactone [74]. One of the two possible chair–chair conformations of this lactone is shown as [80], in which the convex side of the *cis*-decalin system is below and the olefinic groups are somewhat distorted to emphasize the essential chair conformations of the rings. Attack of bromine now proceeds from below on the right-hand olefin followed by intramolecular release of the intermediate bromonium ion by the hydroxyl group to form the ether ([81], R = Br) which locks the decalin into the single conformation shown. This compound need not be isolated, since direct treatment of the reaction mixture with methoxide ion causes elimination of the axial bromide β- to the lactone carbonyl and re-addition of methoxide (axial, from the convex face) to the conjugated olefin thus formed, yielding ([81], R = OCH_3) instead. Writing this product as [75] shows it to have the same stereochemistry as reserpine at all five asymmetric centers around the potential E-ring, obtained in three synthetic steps!

The next stage in the synthesis involves oxidative removal of one carbon from the *cis*-decalin and insertion of the tryptamine nitrogen

CHART I •

The Total Synthesis of Reserpine

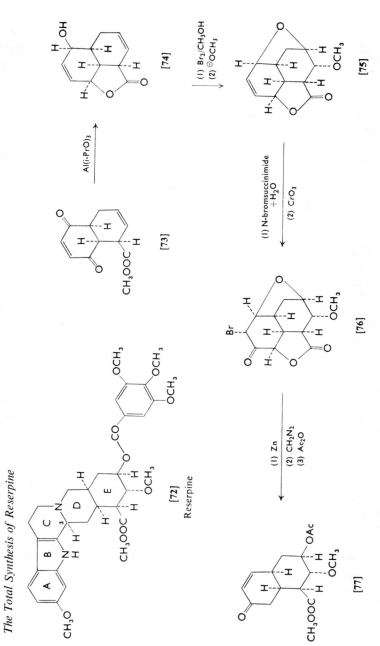

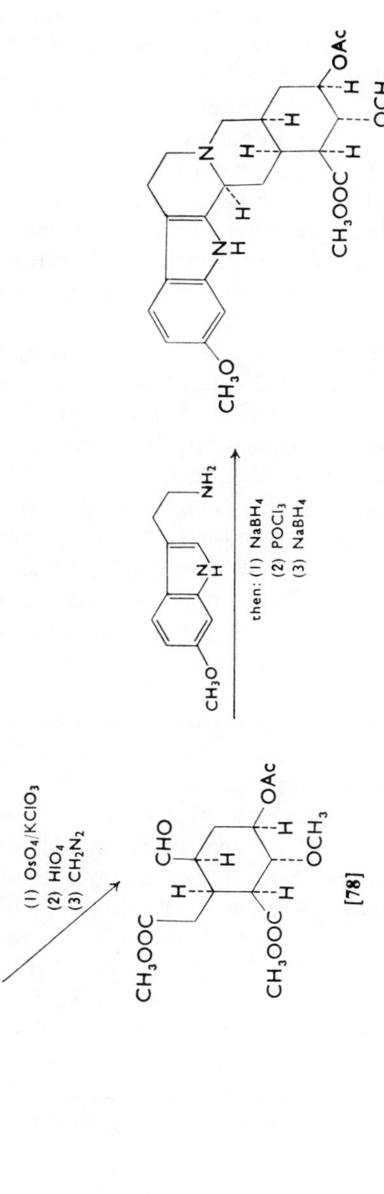

[79]

[82]

[81]

[80]

[78]

then: (1) NaBH₄
(2) POCl₃
(3) NaBH₄

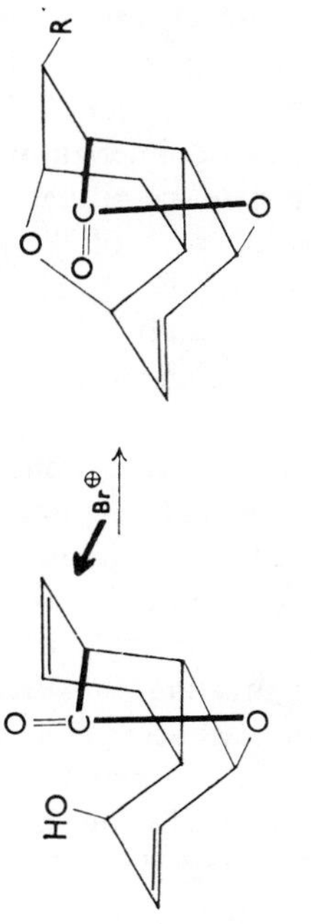

(1) OsO₄/KClO₃
(2) HIO₄
(3) CH₂N₂

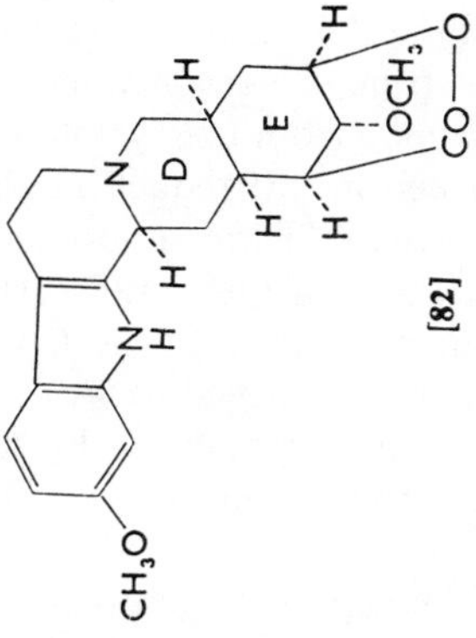

in its place. To achieve this the double bond in [75] was converted first to a bromohydrin (attack of $Br^\oplus$ on the convex face determines its orientation as in [76], and this was oxidized to the corresponding ketone [76]. Zinc, which is used generally for reductive removal of substituents α- to ketones, here expels both the lactone oxygen and the bromide from the cyclohexanone ring while the enolate intermediate formed in this reduction also eliminates the ether bridge, ultimately yielding [77] after esterification (CH_2N_2) and acetylation. Oxidative extrusion of the center carbon of this unsaturated ketone occurs on hydroxylation ($OsO_4/KClO_3$) and HIO_4 cleavage to an aldehyde-acid which is esterified to [78]. This compound is now readily attached to 6-methoxytryptamine by the series shown in Chart 1 (in which no intermediates need be isolated) and yields the epimer [79] of reserpine at C-3.

This epimer is unfortunately the more stable, the indole nucleus occupying an equatorial orientation on the D ring, but conversion of the *cis* equatorial esters on ring E to a lactone [82] locks the *cis*-decalin in the other chair–chair conformation, in which all the groups which were equatorial in [79] become axial. Since C-3 is subject to acid-catalyzed epimerization, it can now be inverted in acid so that the large indole substituent becomes again equatorial to ring D. Then methoxide opening of the lactone yields the alcohol corresponding to reserpine in all stereochemical detail. This complete synthesis is so short and smooth that it is used commercially as a source of reserpine economically competitive with its isolation from the plant!

F. IBOGAINE   The establishment of functional groups in an unknown molecule is usually the first, and easiest, part of the structure elucidation, traditionally inferred from reactions of the groups and, in contemporary practice, to a much greater extent from physical methods. Unraveling the architecture of the hydrocarbon skeleton was a very much more difficult task before the very recent application of NMR spectroscopy. On the other hand, since natural products run to families of compounds with the same skeleton, after the skeleton has been shown for one example, the others are generally settled by interrelation with the first. This process is neatly illustrated in Taylor's economical determination of the structures of the alkaloids from *Tabernanthe iboga*, the source of a popular stimulant in the Congo.

It was clear from spectral and other evidence that the several alkaloids were indoles, with ibogaine, the major constituent ($C_{20}H_{26}N_2O$), being a methoxy-indole. Since the only other functionality was a tertiary amine (not N—CH$_3$) the problem rapidly became one of determining the skeleton. Of the several traditional reagents for fragmenting alkaloids, two were used on ibogaine. The first was KOH fusion which afforded [83] and [84]. It had been separately shown that the methyl group is transferred from oxygen to nitrogen by intermolecular S$_N$2 attack of the indole anion on –OCH$_3$ with displacement of a phenolate ion, thus accounting for these functions in [83]. Taylor observed that these fragments contained all the skeletal atoms if one assumes no overlap and, postulating a tryptamine

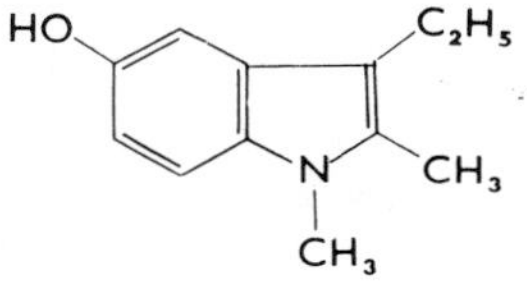

[83] 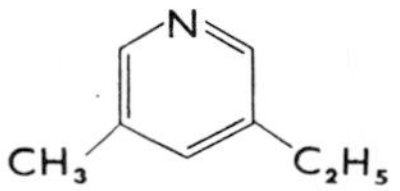   [84]

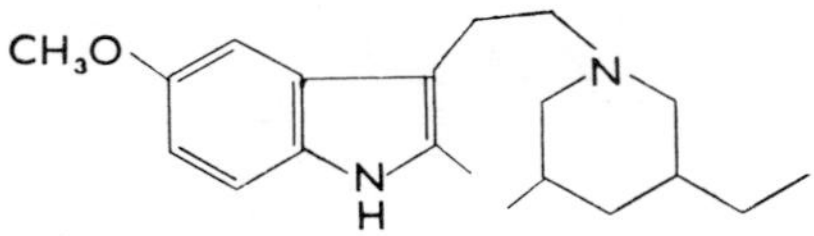

[85]

biosynthetic origin, combined them into the partial skeleton [85], which now requires closure of two rings to formulate ibogaine. Since the ring fused to indole cannot be six membered (without a quaternary nitrogen), it was clear that ibogaine had a skeleton which was hitherto unknown.

The second degradation was selenium dehydrogenation, which effected milder cleavages, yielding a molecule containing all the skeletal atoms. This product (C$_{20}$H$_{22}$N$_2$O) was a secondary amine, since it formed an N-nitroso derivative, and afforded a UV spectrum which was virtually identical to that of the model 2-(o-aminophenyl)-indole. Kuhn-Roth determinations indicated the presence of a C—CH$_3$ and a C—C$_2$H$_5$ (ibogaine itself had been shown in this way to possess a single C-ethyl function). (■) These results account for all of the skeletal atoms of the selenium product except two carbons. However, these carbon atoms can now be placed only as shown in [86], and this has been confirmed by synthesis.

Since the framework of the key product [86] lacks only one ring, only one new link needs to be forged to construct ibogaine and since ibogaine is a tertiary amine that link must be made from the aliphatic nitrogen. It must then join a carbon such as to afford the rings found in the alkali fusion products. Since the only C—CH$_3$ in ibogaine is that of the C-ethyl group, the bond must be made to the asterisked (*)

[86]   [87]
                                    Ibogaine

[88]   [89]   [90]

[91]   [92]

methyl in [86], and this results in the structure [87] for ibogaine, which will now be seen to include the separate skeletons of the KOH products [83] and [84].

For the second part of the deduction, it is necessary to interrelate the other known alkaloids, which appeared to have different aromatic ring substituents. In principle it is necessary to isolate that portion of the molecule bearing the differences (the benzene ring in this case) and show that the other fragment of degradation is the same in all cases. To this end the alkaloids were oxidized to the β-hydroxy-indolenines [88] which rearrange to the indoxyl [89] as shown with arrows. The oxime tosylate [90] of this ketone was prepared and spon-taneously rearranged (see arrows in [90]); hydrolysis of the inter-mediate imine yielded two fragments, the several simple aromatics [91], which easily allowed deduction of the substituent positions in each alkaloid, and the alicyclic amine [92], formed from them all.

G. STRYCHNINE   The alkaloid strychnine [93] was one of the monu-ments of classical natural products chemistry from its isolation in 1817, through the extensive work of Leuchs and Robinson in the first half of the twentieth century, to its final structure proof (1948) and synthe-sis (1954) by Woodward. Most of this work was devoted to oxidative incursions into the underside of the molecule, resulting in clarification of the periphery of related functional groups (shown in boldface in [93]) but leaving the complex center of the molecule largely undefined owing to the absence of any functional groups in this region to serve as handles for degradative entrée. The problem was of course com-plicated by the occasional occurrence of unrecognized rearrangements, and so the course of its history, like that of morphine, was not straight-forward, but strewn with pitfalls and missteps.

A distillation of the essence of this elucidation would, however, include the $KMnO_4$ oxidation of strychnine to strychninonic acid [94]; the recognition of its ketone by reversible reduction to the cor-responding alcohol; and the mild alkali cleavage of [94] to glycolic acid and the nonbasic strychninolone-*a* [95]. This series serves to define the whole outer chain of atoms linking the two nitrogens. Oxidative cleavage of the α-keto-amide appearing in [94] and [95] affords CO_2 and an amino acid which can lactamize, so that this α-keto-amide ring must contain six or more atoms.

The key to the center of the molecule lay in the oxidation of strych-nine to a hydroxy-derivative, pseudostrychnine [93], which now afforded a functional group to lay open the ring system. On the one hand, methylation of this derivative yielded a ketone [96] (R = H) which formed a benzylidene derivative (thus defining an adjacent methylene group). In the related natural alkaloid vomicine ([96], R = OH) this oxidation-methylation has occurred *in vivo* in accord-ance with the biosynthetic reactions of Chapter 2. On the other hand, peracid oxidation of the pseudostrychnine leads to an acyl-indole [97] which was identified by its characteristic UV spectrum and led to the final establishment of the linkages in the architecture of the strychnine molecule.

In view of the low esteem in which KOH fusions are commonly held as meaningful degradation procedures, we may observe that all eight

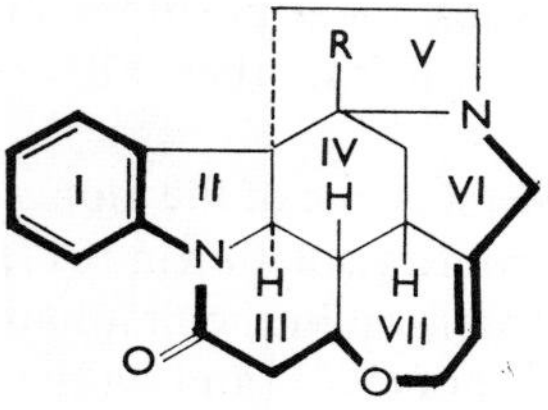

[93]

Strychnine, R = H
Pseudostrychnine, R = OH

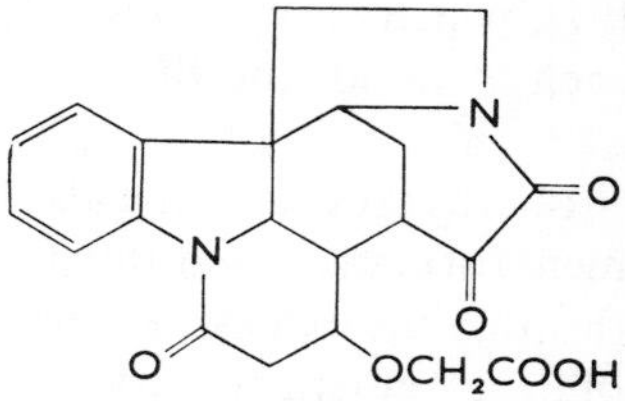

[94]

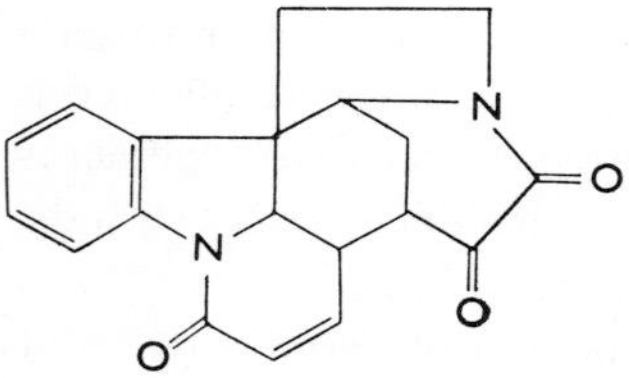

[95]

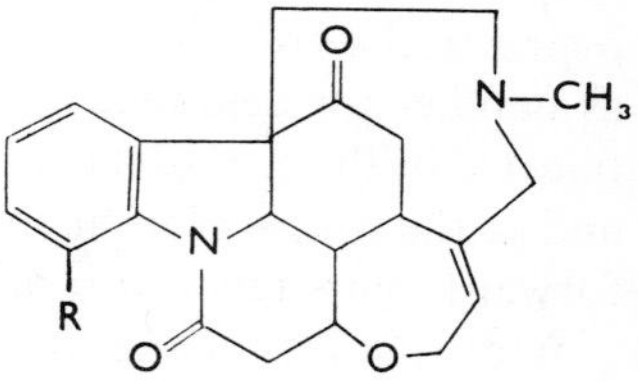

[96]

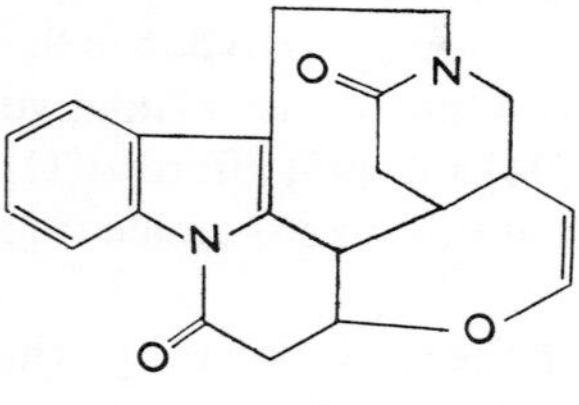

[97]

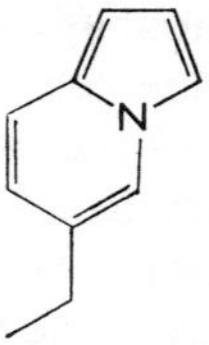

[98]

simple fragments which were early isolated from strychnine by this reaction are found to exist structurally intact in the correct formula [93] of that molecule, while their formation from any of the many older formulas for strychnine required assumption of some skeletal rearrangements in the alkali reaction. The eight products are indole, 3-methyl-indole, 3-ethyl-indole, tryptamine, carbazole (dibenzpyrrole), 3-methyl-pyridine, 3-ethyl-4-methyl-pyridine, and **[98]**. Their formation by successive base-catalyzed reactions on strychnine is, in fact, for the most part mechanistically straightforward!

Although strychnine contains six asymmetric carbons, nevertheless the considerable interlocking of rings in the center actually establishes the stereochemistry without further experiment! (■) Let us start by taking for reference the central ring (IV), in a chair conformation.

[99]

Ring VI must now be fused to it *cis* and hence diaxial and, since common-size rings can never be fused *trans* and diaxial, ring V must also be *cis*-fused to it. This leaves the phenyl attachment an axial one, thus requiring a *cis*-fusion of the indoline ring II also. On the other side, the seven-membered ring VII must by the same strictures be joined *cis*, leaving the juncture of rings III and IV necessarily *trans* and completely defining the stereochemistry as shown in **[99]**; this is actually the enantiomer of the absolute stereochemistry of natural strychnine but makes a clearer picture in relation to **[93]**.

181